Viable Energy Now

When Energy, Economics, and Politics Converge

Jacek Popiel

Published by Roberts & Ross Publishing

Viable Energy Now: When Energy, Economics, and Politics Converge

ISBN: 978-0-9822015-2-7 (paperback)

Printed in the United States

Cover & Interior Design by:
Ronda Taylor

Roberts & Ross Publishing
Englewood, Colorado
(303) 762-1469
Santa Rosa Beach, Florida
(850) 622-5772
www.RobertsRossPublishing.com

TESTIMONIALS

"The energy question increasingly shapes America's foreign and domestic policies—but in the context of "business as usual." In this brief but scintillating book Popiel makes a convincing case for a paradigm shift from growth to efficiency. The fossil fuels whose unprecedented use underpins the world's economic and social order are a finite resource. Systems based on their abundance, capitalism and socialism, are becoming correspondingly obsolete. The challenge is to adapt, and Popiel insists adaptation has been America's historic strength. Popiel's analysis of feasible adjustments in foreign policy and domestic economy, his call for a partnership between state and market, his argument for a political realignment reflecting an efficiency model, can be debated . They cannot be dismissed. This work merits wide circulation and careful reading in both the political and national security communities."

— *Dennis Showalter, Past President, Society for Military History*

"Jacek Popiel, in *Viable Energy Now,* sweepingly explores the global energy business, showing the dynamics and interaction of the business, technology, politics and society. He explores and explains how, in a world with a mixture of socialism and capitalism, we have come to the economy, based on the paradigm of economic growth and globalized financial flows.

As brilliant in the insights as in its elegant, easy readable style, *Viable Energy Now* is a book that should be read by everyone who wonders how we use current technology and strategic partnerships to come from "peak oil" and oil dependence to a sustainable energy scenario, while maintaining economic activity. Popiel shows how the paradigm shift from economic growth to efficiency will result in a "new" economy. He persuasively offers suggestions to the U.S. government, business, political, and military leaders on how to orderly bridge this gap, while allowing for technical innovations needed to secure the future."

— Peter J. A. Tijm

Peter J.A. Tijm is the author of "Gas to Liquids, Fischer-Tropsch Catalysis, Reactors, Products and Process – The Twentieth Century And Beyond" (www.booklanddirect. com). After profound refinery experience Tijm has, for the largest part of his career, been involved in novel energy technologies. He was deeply involved in the development and commercialization of Shell's technologies: coal gasification (Buggenum-The Netherlands) and Fischer-Tropsch (SMDS Bintulu, Malaysia), APCI's syngas technology and demonstration of Rentech's slurry based F-T technology. He is coauthor of more than 70 publications.

CONTENTS

To my wife, Joanna

ACKNOWLEDGMENTS

This book would not have seen the light of day if I had not been privileged to work under Carol and Eddie Sturman, founders of Sturman Industries in Woodland Park, Colorado where I learned much of what I know about energy, innovation and paradigm changes.

I want to thank Mr. Prince Dunn, as well as Dr. Dennis Showalter of Colorado College, for reviewing the original manuscript.

Many thanks also to Gordon Miller and Frank Traditi for their constant support, encouragement and advice.

INTRODUCTION

The primary challenge facing America today is energy. As I write this, the current president of the U.S. has just signed the largest spending bill in our history. It is designed to jump start our failing economy back into its usual pattern of growth: large, expansive, economic growth. But what many fail to understand is that this economic paradigm of infinite growth can only be fueled by an infinite supply of fossil fuels. Very simply, this paradigm works this way: abundant and accessible energy resources allow for indefinite growth of the economy, which in turn supports continually increasing consumption and an ever rising standard of living.

While there is still plenty of oil to go around, the world supply is limited; that is pretty much an accepted fact. That is why there is now a huge push to find "alternative" energy. Wind, solar, waves—all have potential and all are limited in some way. There will be workable alternative energies, but I do not believe that the technology necessary for their creation and manufacture on the scale necessary to sustain our current energy use currently exists. We must recognize that we need an interim solution, a solution that will be able to use the existing energy infrastructure so that we may safely and easily bridge the gap between fossil fuels and true alternative energy sources. That is what this book presents—a viable energy alternative that can sweep us into the next era of energy use.

Abundant and inexpensive energy, the essential ingredient for economic growth, is what fueled the Industrial Revolution, the moment in history when the concept of infinite economic growth appeared. This energy has been provided, of course, by fossil fuels. First came coal, and on its heels, petroleum and natural gas. The territory of the United States has been endowed with an abundance of all three. It is, therefore, no surprise that the U.S. has, more than any other nation, embraced the paradigm of continuous economic growth born of the Industrial Revolution. Here's just one example of the massive impact America's use of fossil fuels has created: just as coal was the foundation of modern manufacturing, oil has been the fuel of transportation and trade, without which the modern, globalized economy is unconceivable. Once the world's foremost oil producer, the U.S. pioneered mass automotive and air transportation, and the rest of the world followed its lead, until nations and continents became connected by a huge network of trade and transportation channels.

Fossil fuel reserves, although huge, are nevertheless finite. The nation is now importing two thirds of the oil it uses, and the ratio of imports to domestic production continues to rise. The oil price spike of 2007 – 2008 was the warning bell that global growth was pushing against the limitations of world oil supply. While the economic downturn has provided temporary relief, the oil supply is not increasing: older fields are still declining while new projects are being delayed or canceled. Should global growth resume, then high prices, aggravated by speculation, will rapidly return, precipitating further crises and economic downturns.

In the minds of many, "renewable" energy sources must now replace fossil fuels. None of the currently known sources, mainly solar energy, wind and bio-fuels, can compete with the huge advantages of petroleum in terms of flexibility, ease of use, and established infrastructure. There is, at present, no substitute for oil.

There is a viable interim solution. It is based on the known fact that U.S coal reserves will last for generations, but it is also inescapably political because when energy and the economy meet, it is *de facto* political. This solution, in fact has two main components:

The first is a change in attitude. We cannot *grow* our way out of our current predicament. Growth is not a solution. It is, in fact, the problem. Unlimited economic growth requires abundant and inexpensive energy, a pre-condition which can no longer be safely assumed. We must thus begin to turn away from the universally accepted paradigm of economic growth and toward the new paradigm of *efficiency of energy usage.*

The second component is to ensure the security and volume of our oil supply until such time as acceptable substitute sources of energy are developed and put in place. Two thirds of our petroleum usage is for transportation, something we cannot, at this time, do without. In other words, we need a secure source of fuel. There are only two such sources currently available: petroleum products and synthetic transportation fuel made from coal and natural gas, both of which the U.S. has in abundance. They are fully interchangeable and entirely compatible with the existing energy infrastructure.

Putting both components in place requires a third one: political will. Here the American political tradition provides the U.S. with a significant advantage.

For two centuries the industrial world has been internally divided by the struggle between two adversarial ideologies—capitalism and socialism. The debate and attending social upheavals has been essentially over the disposal of the wealth created by the Industrial Revolution, itself founded on the use of fossil energy. Both ideologies are incomplete and one-sided: socialism idealizes the state, while capitalism assigns all knowledge and wisdom to the markets. Neither has succeeded in its pure form, and all states have in the end adopted hybrid systems that blend both in varying degrees.

This blending, while it has softened social conflict and produced relative material abundance, has nevertheless major disadvantages. It provides neither the full economic efficiency of capitalism nor the total social fairness of socialism. The ideological divide between Left and Right still continues within the economy and the government, creating confusion over the issues and preventing the elaboration of strategies on which a national consensus can be based.

Such deadlocks have occurred in past U.S. history, particularly in the years preceding the Civil War and in the decade of the Great Depression. In each case the nation was faced with major policy choices, and each time it responded, developing practical responses that trumped pure ideology. I believe that ours is just such a time, and that a similar national consensus can be created to answer the current challenge.

The solution will require us to put realism above political or economic theory, and national unity above party and factional interests, something the U.S. has always been good at, particularly in emergency situations. A new political compact will be needed, transcending the current positions. New political alignments, possibly a new political formation, will emerge, responding to the challenge.

This book is an attempt to better define this challenge, and to propose a platform around which the national discussion could start.

COAL, OIL ... AND THEN WHAT?

The origins of our current energy situation lie back two centuries, at the beginning of the machine age. The Industrial Revolution, powered by fossil fuel, produced the concept of indefinite growth, which still guides our economic policies. It also gave rise to the competing ideologies of capitalism and socialism, which have dominated the political sphere.

Now as our key fuel, oil, is becoming less available and more costly, this entire edifice has to be restructured.

CHAPTER ONE

The Paradigm of Infinite Growth & the Shift in Energy Availability

Energy is the foundation of the modern economy, and for the past century and a half "energy" has primarily meant *fossil energy*: coal, oil and natural gas. A growing and inexpensive supply of all three has supported continuous economic growth for so long that such growth has become a given.

This situation is changing. While abundant reserves of fossil fuel still exist, in the critical area of petroleum our ability to expand supply at a reasonable cost is reaching its limits. Our economic culture is approaching a fundamental inflexion point, requiring major decisions. To understand how important this point is we need to go back in history to the beginning of the machine age, or the great economic upheaval known as the Industrial Revolution.

Our current economy and material way of life are rooted in this change, which began during the eighteenth century, and gained full force during the nineteenth. Up to roughly 1750 all human economic activity had been strictly of the "sustainable" kind. This was not by choice but by necessity. All available inputs used in economic activity, such as forests, fields, and domesticated farm animals, were what we would call today renewable. The only power used was human or animal, with water and wind added whenever feasible. Most of the available energy was spent to simply to keep people and their domestic animals alive. Whatever surplus remained after that was invested in durable assets: buildings, fortifications, mills, and a range of high-quality goods, such as tools and arms, produced by skilled and experienced craftsmen. The priority given to such long-term investment was the natural response to the vagaries of weather, poor crops, wars, and plagues. With luck, such durable assets would outlast times of trouble and slowly accumulate, contributing to a gradual rise in the standard of living. This is what occurred in Europe between the ninth and the seventeenth centuries, leaving behind a legacy of fertile fields, lasting buildings, and magnificent art.

This age-old pattern of slow wealth accumulation radically changed in the 1700s through the introduction and development of mechanized manufacturing. The use of machines allowed a quantum leap in the amount of goods produced and by the same token lowered their price. Advances in engineering and the mechanical crafts, as well as in finance, were critical to this development. But the fundamental difference with the past, and the major change which made the Industrial Revolution possible, was *the use of fossil fuels*. These provided the extra energy to power the new mechanized industries and the associated means of transportation. Without the introduction of this new source of power the large-scale use of machines would have been impossible.

The use of fossil fuel had two advantages: first, these were available in apparently limitless quantities, thereby eliminating the constraints associated with the renewable power sources that had been used before. If, for example, wood is used as a fuel, it first needs to grow, and the rate at which this takes place limits the amount of wood that can be harvested in any given year. Coal, on the other hand, can be dug up at any rate one chooses. Second, fossil fuel delivered far more energy than was needed to procure it. It is this *energy surplus* that in the end made mechanization possible, and with it the vast increase in material production brought by the Industrial Revolution. The more accessible the coal deposits, the greater the amount of disposable energy the extracted coal provides, and the greater the rate of growth of the associated industrial development.

In the course of the second half of the eighteenth century, the constraints that had, from time immemorial, both shaped and limited material development were thus removed. Everything that had been laboriously produced by hand could now be manufactured much faster and in larger quantities by machines that were in turn powered by fossil energy. Whereas in the past, economic development meant the slow accumulation of high-value, durable assets, it now gradually shifted to the rapid production of goods that could be consumed. *Growth*, rather than accumulated wealth, became the new paradigm of economic success. The apparently limitless availability of fossil-based physical energy allowed the economy to expand significantly within a matter of years, rather than generations as had previously been the case. Growth meant higher incomes for the producers and greater comfort for the consumers. For governments it meant a constant expansion of tax revenue with which armies could be equipped and empires acquired. Growth was desirable for all concerned and became the paramount goal of all economic activity.

Increased production, to be profitable, required both greater and faster consumption. So the logic of growth for its own sake led to the fostering of the mass market and of the consumer economy. The *economic growth* paradigm demands that every individual constantly increases his or her consumption of the goods produced. Within this economic model even waste, as in the case of disposable containers and packaging, is useful, since it contributes to growth. The value, quality, and usefulness of goods produced also lose significance. The growth paradigm of the modern era is thus the complete opposite to that of the previous historical periods, which favored the slow accumulation of goods having both great value and long life.

The fact that goods are of lesser quality and more transient value does not matter under the growth paradigm, but this only holds true if physical energy is both available and inexpensive. And that is the way it has been for approximately two centuries, since the economic growth paradigm became established in the late 1700s. The nineteenth century saw a vast expansion of industrial production, based primarily on the coal-fired steam engine. This expansion continued during the twentieth century, with the emphasis shifting to the internal combustion engine, and later the turbine, both powered by fuels associated with or derived from petroleum.

Petroleum is the king of fossil fuels. It has a high energy content and often flows spontaneously out of the ground. It can be stored at room temperature, cheaply transported by tanker or pipeline, and safely handled. It is available in huge quantities and until recently new reserves have come on line as fast as the older ones became exhausted. In addition to its use as a fuel, oil serves as raw material for a variety of chemical processes, just as coal did before. Today's transportation infrastructure is almost entirely based on petroleum, with the United States (once the world's greatest oil producer) in the lead as the most mobile nation in the world.

In summary, the discovery and large-scale exploitation of fossil energy have been, and still are, the foundation of the modern economy. Fossil energy shapes both our material way of life and the way we think about it. Its existence, availability, and use underlie all commonly accepted assumptions of economic policy, including the primary one that the Gross Domestic Product must expand indefinitely, as must individual and collective consumption. Because fossil fuel has been available in such extraordinary quantities, these assumptions have never been questioned. Until now, that is.

THE DAWNING LIMITATIONS

As stated above, all pre-industrial economic systems were, by necessity, "sustainable" and based on renewable resources. They could not use such resources at too fast a rate without destroying their economic foundation. Some civilizations did overreach in their use of resources, and perished as a result. But even when a state or empire drifted into this form of gradual economic suicide, it took decades, if not generations, for resources such as forests or fertile topsoil to become exhausted. To the populations concerned, these resources appeared so large as to seem inexhaustible, and their excessive exploitation did not appear to have destructive effects until close to the end, when it became evident that something was amiss, and it was too late to fix it.

We are in a somewhat similar situation with respect to fossil fuels and the various forms of energy derived from them, except that we still have a choice. The reserves accumulated during a good part of earth's geologic history are indeed enormous, and the progenitors of the Industrial Revolution were excusable when they considered them to be, for all practical purposes, inexhaustible. But it was impossible for them to predict that the new industries they had created would become the foundation of the economic growth paradigm, with consumption growing exponentially while the total amount of fossil energy resources did not.

In the end everything on the globe is finite, including fossil energy reserves. Economic growth as understood today, on the contrary, is assumed to go on indefinitely. This is not feasible when resources, however abundant, are still finite in the end. Sooner or later we will run into some kind of limitation, and this is now happening with our most valuable energy resource, petroleum.

In the energy industry, the point at which existing petroleum reserves begin to decline faster than new ones come on line is termed "peak oil." There is considerable discussion about whether we have reached this point: the maximum production level that will ever be feasible. Beyond the "peak oil" point increased consumption of petroleum is no longer possible. It will plateau for a while and then start decreasing. There are good reasons to believe that "peak oil" is not far away. Some in fact place it in this decade.

The peak oil point is not an absolute. It will be affected by a number of variables, such as the probability of new discoveries, the development of recovery techniques, the improved ability to locate deposits, the effects of economic conditions on demand, the response to higher prices, and so on. Two additional concepts must be taken into consideration.

The first is what can be termed the *energy margin*. This refers to the percentage of energy left in a barrel of oil after the total energy needed to extract it has been subtracted. For example, many of the early oil discoveries in the USA were shallow deposits that flowed under their own pressure, creating "gushers." Deposits being developed today may be in deep offshore waters and require massive drilling and pumping installations, or in hard rock that must be broken up for the oil to be extracted. Slowly but surely the energy margin is shrinking.

The other factor is simply cost. Even if containing an abundance of petroleum, a hard-to-exploit deposit will result in expensive oil. Cost is still low in a number of oil producing areas such as Kuwait, but even a cursory analysis of new fields now coming on line, such as the Kashagan field in the Caspian, the Khurais field in Saudi Arabia, and the new "pre-salt" discovery off the Brazil coast, shows that the cost of extraction is inexorably rising.

Both the rising cost and gradually decreasing physical availability of petroleum undermine the indefinite growth paradigm in place since the beginning of the Industrial Revolution. While there are, at this point, no such limits for the other major fossil fuel, namely coal, the gradually tightening constraints on petroleum price and availability will require substantial changes in the way we use and think about the role of energy in our economy.

The 2007–2008 oil-price spike was a striking demonstration of the importance of oil supply for the global economy. While the spike was sharpened by speculation, it was based on a real imbalance between demand and supply. While the price has dropped due to the recession that followed it, the next spike is already building up.

As stated before, petroleum is the king of fuels. There is no substitute for it, as will be explained in some detail later in this book. Its gradual fading from the energy scene will cause a cascade of adjustments, and will, in the end, require the development of a new paradigm for economic activity. Defining this paradigm will be one of the major challenges faced by society in the twenty-first century.

The formulation of this alternative outlook and approach will take time. It will be a collective undertaking, just as the Industrial Revolution was a collective endeavor that required a couple of generations to pass before its results could be seen and understood. The emergence of a new paradigm will have major consequences, for not only our economy, but also our social organization and our political systems, which have been profoundly affected by the Industrial Revolution. It is to this political aspect that we will turn next.

CHAPTER TWO

The Old Political Paradigm

THE CONTROL OF THE MEANS OF PRODUCTION

Fossil energy was, and still is, the central foundation of the industrial age and the modern economy. Its abundance underlies the concept of constant economic growth, which all modern economists and consumers consider a given. But the influence of energy goes beyond the purely material: the Industrial Revolution not only generated unprecedented abundance, but also a *new political paradigm dealing with the disposition of such abundance.* This new paradigm is the *divide and opposition between Right and Left, between capitalism and socialism.* To understand how this opposition arose we must again go back in history.

We have already stated that, before the changes brought about by the availability of energy from fossil sources, economic development was based on gradual wealth accumulation. Political stability was a prime requirement for such accumulation to be possible, so the necessary counterpart to the sustainable economy of the pre-industrial age was a political system that was static or evolved very slowly. War or political upheaval on the modern scale, such as revolutions or total war involving mass mobilization, would have been suicidal in those times: they would have quickly exhausted the economic resources available. Therefore, conflicts tended to be settled within the existing political structure under rules aimed at minimizing expense and collateral damage. When the rules broke down, as in the case of the Thirty Years War in Germany (1618–1648) the resulting devastation took generations to repair.

The pre-industrial political system was an interdependent structure within which the leading components of society, the "orders" of aristocracy, bourgeoisie and church, were balanced against each other and at the same time bound by a set of mutual duties and obligations. This structure, elaborated by trial and error during the Middle-Ages, was designed for stability and held together by a set of beliefs and rules assumed to be heaven-ordained, eventually culminating in the concept of hereditary monarchy by divine right. It was a coherent and eminently practical system, and over roughly a millennium it allowed the nations of Western Europe to rise from the status of barbarian backwaters to that of ruling global powers.

This elaborate structure of duties and privileges was thrown out of balance by the sudden increase of wealth created by the Industrial Revolution. With it a new social component was born out of the merchant class: the industrial capitalists. Within the space of a couple of generations the wealth of this social group surpassed anything that had, up to then, been thought possible.

Simultaneously, the new machine age brought with it a belief in continuing human progress, ushering in a new golden age of science and prosperity, clearly separating it from earlier periods of history. How this new era would be managed became not only a practical, but a philosophical issue as well.

Since the new material abundance was based on the mechanized factories where the flow of goods originated, the key question centered on how and by whom these facilities would be owned and managed. To phrase it in the words used at the time, the main political debate arising from the Industrial Revolution was over *the control of the means of production.* The answers given to that question have defined the political alignments of the last two centuries. And they still do, though most of us are no longer aware of it.

The sudden potential to provide apparently infinite wealth did not fit within the old system based on gradual evolution and slow accumulation of valuable assets. There was no question, as the Industrial Revolution began to unfold, that a new way of dealing with the distribution of wealth had to be designed and a new political structure set up to manage it.

Two schools of thought arose out of this quandary. The first, which eventually became the conservative approach, was to leave control to the creators of the new abundance: the industrial capitalists. Since this group, through talent, willingness to take risks, and creative management, had brought about this increase in human fortunes, they should be entitled not only to lead it into the future, but also to enjoy its fruits. This meant that the bulk of the new wealth being created should remain the property of the capitalist class. This wealth would be best left in the hands who had initially made its creation possible, and who were, therefore, most qualified to make it grow further.

The second approach took the opposite position. The new wealth, other thinkers argued, was produced not by the industrialists themselves but by the masses of workers who toiled, often in abject conditions, in the newly created production facilities. Allowing the industrialists and bankers to keep and enjoy the fruits of this labor was unjust. It would also be wasteful, since, once rich, the capitalists would spend their immense profits on useless luxuries, rather than on the general improvement of mankind. The social order that allowed such abuse had, therefore, to be overturned. The exploited masses had to be organized politically, and the reactionary governments colluding with the capitalists overthrown by revolutionary action. The new political structure emerging from this process would then take over and ensure fair distribution to all. This was the core of the alternative doctrine, which came to be known as socialism.

This fundamental dichotomy, the division of political thought in two opposed and inimical camps of capitalism and socialism, became the dominating issue within western political life. The split was originally about economics, and at the core of the debate was the disposition of the new industrial economy and of the wealth it produced. Without this wealth the debate would not have taken place. As the key foundation of the new industrial system was fossil energy, the new political alignments owed their existence to fossil energy as well. Take energy out of the historical equation, and capitalism and socialism as we know them cease to exist.

THE LEFT AGAINST THE RIGHT

It took roughly one century for the two camps to develop their respective ideologies and to get organized politically. The struggle between the socialists and the capitalists, between the defenders of the free market and the proponents of state power, gradually became the main reason for existence of the diverse revolutionary and reform movements that dominated the political life of the nineteenth century. In this contest of ideologies both sides appropriated elements of earlier political debates: the structure of government, the issues connected with human rights, the role of religion and culture in political life. These issues were then incorporated into the two opposing world outlooks. The nineteenth century thus forged the ideologies around which the wars and struggles of the twentieth century would be fought.

Both sides firmly believed they had right on their side, so the contest gradually became an all-or-nothing struggle. Ideological positions hardened, and compromise as a means of peaceful resolution was largely abandoned. As the nineteenth century progressed, the divide between Left and Right widened, with each side attempting to crush and destroy the other. Over time these two ideologies replaced nationalism as the main political motivators. The failed revolution of 1905 in Russia was the opening battle of a contest that would last until the end of the century.

The struggle continued, with ups and downs for each side until 1949, when China fell into the socialist camp. The world was now roughly divided between two great alliances, one of socialist states, the other of capitalist ones. Both were armed with nuclear weapons and capable of fielding massive armies. On the socialist side every human activity was under the control of the state. On the capitalist side, the free market system was preserved. The socialists had full control and the power flowing from it; the capitalists retained private initiative and its huge advantage in terms of economic efficiency.

The resulting stand-off was the Cold War. Within the internal logic of each side it was considered necessary and inevitable. Seen from a wider perspective it could be considered a massive waste. Enormous energy and resources were expended in a military and economic stalemate that should never have existed. It might have been otherwise had the dependence of the industrial economy on the consumption of fossil energy been better understood. The myth, however, of the limitless riches it offered, together with the promise of a new and final era of endless prosperity, was too attractive and powerful to allow for rational analysis and political compromise.

The Cold War ended with the collapse of the Soviet Union as a political entity, and the capitalist system was declared the victor. "Free markets" were declared to be the true answer and the superior formula, and capitalism was given, just as at its beginnings, free reign to expand, enrich its practitioners, and solve mankind's problems.

There was a two-fold error in this conclusion.

First, the triumph and worldwide spread of the consumer culture as developed in the West only exacerbated and brought forward the inevitable clash between finite energy resources and the promise of indefinite economic growth. Second, it was not the wild and unbridled capitalism of the 1800s that had won the victory. The "victor" was, in fact, a highly socialized political and economic system, where the power and reach of the state had grown enormously, and industrial activity was hemmed in by numerous constraints involving labor conditions, social safety nets, diverse entitlements, environmental regulation, private and corporate taxes, and so on. Capitalism had not in fact defeated socialism, nor had socialism vanquished capitalism. They had *blended*.

Each state or federation of states now had its own hybrid system, differing only in matters of degree. In the midst of the Russian Revolution, Lenin had already come to realize that the state could not control every nook and cranny of the economy. As a result he had set up the New Economic Policy, which allowed for a limited amount of private enterprise. Stalin quashed this policy, in the process condemning the Soviet economy to perpetual inefficiency, but private enterprise never entirely disappeared from the Soviet system. In fact it grew and spread, always unofficially, as the economy became more complex and the limitations of central planning were increasingly exposed. There were private plots and domestic animals, open markets and hard currency stores, industrial fixers alleviating shortages, underground businesses, and all kinds of off-plan initiatives. Unable to control these developments, later Soviet leaders such as Andropov and Gorbachev realized that they might as well make room for some form of capitalist efficiency. They, therefore, attempted to modify the socialist system accordingly, just as other communist states such as Hungary, Poland and Yugoslavia had already done.

A similar assimilation process had been at work in the West since the late 1800s. Labor unrest on the one hand, and simple humanity on the other, convinced both industrialists and government officials that workers could not be treated like abused pack animals. This trend was reinforced, after 1917, by the threat of communist subversion; and, after the crash of 1929, by the evident failures of laissez-faire capitalism. Even though the resulting reforms, were initially decried as Marxism in disguise they did promote social harmony and weakened revolutionary trends.

More importantly from the business point of view, they enriched the working class. This rise in prosperity eventually resulted in the creation of an increased number of consumers and the growth of new markets. While entrepreneurs and business executives attended to the new opportunities, politicians came to realize that offering various government-funded benefits was a handy way of securing votes. Whatever purists might say, and however evil socialism might be in theory, its milder forms sell very well in practice.

The struggle between Left and Right, between conservatives and progressives, has been played out over the entire globe, with each nation or group of nations being affected somewhat differently. The United States has its own specific history in this regard, which we will now review.

CHAPTER THREE

Capitalism vs. Socialism in America

Although it has been spared its most violent manifestations, the U.S. political system has not escaped the divide between left and right. Our two main parties have each identified with one side, the Democrats with the Left and the Republicans with the Right. But the gradual blending of capitalism and socialism has slowly erased the differences between them, leading to the current political stagnation. Forward progress can resume only when a new and current set of issues are defined and party alignments adjusted accordingly. Possibly an entirely new political formation will be needed.

The development of modern industry occurred simultaneously with the birth and growth of the political system known as representative democracy. As western society went from being primarily agricultural and rural to mostly industrial and urban, representative democracy made parallel and nearly simultaneous strides. Thus, the political development of the modern western world, the main feature of which was the development of democracy, has also to a great extent been dominated by the issues created by the industrial revolution.

Modern democracy was nurtured in England and the United States of America, and within those two states it developed its most stable form, based on the two-party system. The great advantage of this arrangement is that, unlike multi-party systems, it allows for the easy formation of a stable majority with a clear mandate to rule. At the same time, however, it also enables the opposition party to gain power through a relatively small shift in voter preferences, thus maintaining political balance. The two-party system avoids the complexities, compromises, and potential paralysis of multi-party coalitions, and yet is remarkably responsive to changing patterns and preferences within the electorate. It has proven itself to be the most practical and enduring form of representative democracy.

Within this system a political party needs a distinctive platform in order to attract voters and to support discipline and commitment among the registered party members. Such a platform in turns needs to be founded on a clear set of philosophical principles, around which the voters' interests can be organized into a coherent whole and then translated into policy. Within a two-party political structure there must be two such sets of principles, different enough to allow the voters to distinguish between parties, but not to the point of being mutually exclusive—in which case the political structure would split into hostile halves, as happened in the U.S. on the eve of the Civil War.

The Civil War settled the basic issues connected with the nature of the U.S. political system: national unity and the respective powers of the states and the federal government. Once the war was over it was inevitable that the political parties would define themselves according to the great questions raised by the industrialization of the country. Such definition inevitably entailed a choice between Left and Right. This alignment took time to crystallize, because up to roughly 1870 the key preoccupation of the nation was with westward expansion. It was only after the Frontier had reached the Pacific that full attention could be given to economic growth and the control of the means of industrial production, which were by then replacing land and agriculture as the main source of employment and wealth.

America's physical separation from Europe tended to mute the ideological intensity of the national debate between Right and Left, but the same issues would nevertheless have to be dealt with. Between 1865 and the turn of the century both American political parties borrowed ideas from the Left and Right simultaneously and interchangeably. After the presidential election of 1912 their paths were set: the Democrats became identified with the "progressive" approach (closer to the Left), while Republicans took the "conservative" side (leaning to the Right).

These orientations have clearly defined the two parties in the public mind since the New Deal. The resulting perceptions were further reinforced by the policies of later presidents, particularly those of Lyndon Johnson and Ronald Reagan. Even today many think of the Democrats as "the party of the people" and of the Republicans as "the party of business." And for a long time their respective platforms have been shaped accordingly.

But as capitalism and socialism have blended, the differences between the parties faded. In order to maintain their distinctiveness and their appeal, both parties have looked for, and embraced, new causes that were compatible with the interests and leanings of their respective voter bases. The Republicans thus embraced "traditional" values while the Democrats went for "diversity" and "tolerance." The parties also sought to recruit newly defined fractions of the electorate such as women, Hispanics, urban professionals or blue-collar males, whose views were more or less compatible with each party's fundamental orientation. None of the new causes, however, could generate the dedication, interest, and intensity of the old Left-Right struggle, as it peaked during the Cold War. American survival against the communist menace was worth going to war for. "Diversity" or "free markets" are not.

At the same time, the addition to the party base of new groups or factions, while providing a numerical gain, also spread out and diluted the party platform. Having thus lost over time their original ideological center and distinctiveness, the two national parties have gradually morphed from committed and disciplined organizations into amalgams of disparate lobbies and interest groups.

Many of the negative features of today's politics derive from this loss of political core. When there is no longer a higher cause to serve, dedication and service can no longer be counted on. They must be purchased, hence the growth of the power of lobbies and interest groups. When there is no clear party identity, one must rely increasingly on an artificially created image, positive for oneself, negative for the opponent. This leads to an ever growing use of manipulative advertising, with the associated escalation in election campaign costs. With no overriding purpose to achieve, political activity becomes dominated by self-preservation and incumbency, even if that means satisfying every interest group in sight at the expense of the nation as a whole. At the end of this dilution process, politics becomes nothing but a game played by professional operatives eager to hire out to the highest bidder.

The most telling signs of this political dissolution are the ever lower approval ratings of the governing institutions and, more importantly, the growing numbers of independent or uncommitted voters. There was a time, not so long ago, when one would be either a Republican or a Democrat, and that was it. Switching allegiances was rare and considered a mild form of treason. Being "independent" was to most voters an incomprehensible concept. There could be no point to it because the key issues were clearly defined. One had to be on one side or the other, and there was no middle.

With the political elite preoccupied with its own self-preservation on the one hand, and on the other an ever increasing number of disenchanted voters sitting on their hands, the political machine is now close to being both directionless and paralyzed. In a time of peace and abundance, such stagnation can be tolerated. It cannot be when there are major problems demanding a solution. At the fundamental level there are two major issues.

The first problem derives from the contradictions inherent in the blending of capitalism and socialism within a single political and economic envelope. Such a merger does not necessarily mean reconciliation or integration. In this case it has been a compromise of convenience. Hard-core members of both the Left and the Right are still convinced of the superiority of their respective ideologies, although they are no longer willing to die for them.

Conservatives have interpreted the collapse of the Soviet block as a license to return to unbridled "free markets." But markets are no longer free: the old industrial economy is now shackled by numerous government-imposed constraints and controls. The tentative conservative solution to this dilemma is the escape into a globalized economy where capital and goods can move, at least in theory, without interference from governments. But the power and reach of states and governments have grown, and with that their ability to tilt the economic playing field according to national interests. As a result, globalization has not created a free world market but instead a confused arena where the free market policies of some states mesh with the blatantly nationalist practices of others. These differences have led to major economic imbalances without a supreme arbiter to set them right. The resulting economic and financial collapse has put the very concept of globalization in serious jeopardy.

On the "progressive" side, the general acceptance of at least some socialist principles has saddled the governments with ever greater obligations without giving them the necessary resources. These resources are greatly affected by an uncontrolled global economy even as governments are requested to do ever more for their respective populations. At some point something has to give. For the U.S. right now, the weak point is the currency. Excessive and continually increasing trade and budget deficits are creating a situation where a run on the dollar, be it sudden or gradual, would have the potential to collapse both the U.S. economy and the global financial systems.

Up to now the solution of the above problems has been economic growth, which provides ever more government revenue, thus allowing the state to provide more services. We have stated that the growing restrictions on petroleum supply are ruling this solution out. Never mind, the conservatives would answer, if demand is there the market will manage to provide an adequate supply of oil.

The market, however, is no longer the only player in town. Energy supply is no longer a purely economic matter, but has been invaded by strategy and politics, as is explained in the next chapter.

CHAPTER FOUR

Who Owns the Oil?

We have outlined above how the exploitation of fossil energy has provided the foundation for both the economy and the politics of the industrial age. We have also pointed out how the most flexible and widely used of these resources, petroleum, is becoming limited in terms of both availability and affordability. The fundamental factor here is the fact that older deposits are running out and new ones are harder to reach and costlier to exploit. These physical limitations, however, are not the only constraint on the oil supply. They are reinforced and amplified by the profound change that has come over the oil market: the rise of energy strategy and politics.

One of the enduring legacies of socialism, despite the "victory" of free market capitalism in the 1990s, has been the growth of both the reach and power of the state. No government today intends to withdraw its influence from the economic sphere. The vast majority of states manipulate economic trends to their national advantage, whatever lip service might be paid to "free markets." This trend is growing, not fading away, a reality that flies in the face of all globalization theories.

The temptation for governments to increase their economic influence is further enhanced by the consequences of the clash, already referred to above, between finite resources and accelerating economic growth, a clash that the process of economic globalization has brought forward in time and exacerbated. The result of this collision between growth and resource limitations, between concept and reality, is the phenomenal rise in raw material costs witnessed over the last decade, the very years during which globalization has appeared triumphant.

Governments are, of course, interested in economic growth. But their first priorities are in the fields of finances and national security, or, to put it bluntly, taxes and guns. No government can operate without a budget, and a major part of that budget will always be allocated to the maintenance, and often the expansion, of military forces. In a time of intense economic competition such as globalization has fostered, the value of military strength rises considerably, as is evidenced by the rise of the military expenditures of great powers such as the USA, China and Russia.

The petroleum industry fits right within the above trends. The oil business was initially developed by entrepreneurs, and later taken over by oil companies, which in the process acquired extraordinary expertise and know-how relating to locating and exploiting oil deposits. For these corporations, oil was a product to be extracted, processed, and sold. In order to grow the business and supply their customers, they looked for oil wherever it could be found. Through the hazards of geology, this was often in states that were politically weak and considered backward by westerners. The modus operandi of the oil companies was to buy an exploitation license from whatever government was there, pump the oil out, and pay the state a limited royalty. They had no interest in raising the price of the product as long as they made a decent profit. In fact, the oil majors long kept the price low in order to expand their market.

While some institutions, such as the British Navy, early on became aware of petroleum's military value, it was only after WWII that its strategic importance became clear. As the European colonial empires collapsed between 1945 and 1970 the vast majority of their colonies became independent and acquired a new political consciousness. From then on it was only a matter of time until the oil-producing states realized that the petroleum deposits within their borders were much more than just a useful commodity.

The years following colonial emancipation saw not only the first major rises in oil prices but also the gradual nationalization of oil deposits worldwide. Today between 80 and 90 percent of world oil reserves are in the hands of state-owned corporations, with the oil majors being progressively reduced to the role of exploration and drilling contractors. While the recent takeover of the Russian oil industry by the state has generated many negative comments in the western press, the fact remains that Russia only did what the vast majority of oil producing states, including U.S. allies such as Saudi Arabia, had already done years ago. And since the most promising areas left to explore for oil and gas are within the countries where energy is already under state control, the trend toward state ownership of oil fields is irreversible.

This trend represents more than a simple change in ownership. The attitude of governments toward energy resources differs from that of private corporations on two major counts. First, states do not look on oil and natural gas revenues as a source of profits to be distributed to shareholders, but as a source of cash to finance the national budget. For this reason they will drive the price to the highest level the market will bear, and beyond that if feasible. Since it is now becoming clear that oil is becoming a scarcer resource, they have no interest in lowering the price so as to broaden the market. The higher the price, the fuller government coffers will be, and that is the state's first priority.

Second, governments think more strategically than corporations do. They not only have, in general, a much longer planning horizon, but in addition, when setting policy, they take into account political and strategic considerations and not just economic ones. Energy resources, therefore, once in the hands of the state, will inevitably become a strategic tool. Russia recently has been, quite correctly, accused of using its oil and natural gas resources in such a fashion. But every state in Russia's position will do so, maybe less blatantly, but with the same desired results in mind.

The governments of oil-producing nations will not, of course, kill the goose laying the golden eggs. They will not raise the price and/or reduce supply to the point where the market collapses. But all the incentives are on the side of keeping the market tight and the price high, at least until a workable substitute for oil is developed or discovered. But as such a development is still far away, the situation of relative scarcity is likely to remain the norm. The current policies of oil producers are a good illustration. Under the "free market" theory they should increase supply to lower the price and stimulate demand, but nothing of the sort is happening. Instead production is sharply curtailed to bring prices back up, just as suggested above.

State ownership of resources does not fundamentally change the way energy supply influences economic activity but it complicates it, since politics as well as economics and technology must now be taken into consideration. In a theoretical free market all customers are equal, and availability and price are, again theoretically, determined only by supply and demand. The petroleum market is no longer even remotely free, ideal, or theoretical. It has been heavily invaded by politics, which only adds urgency to the energy issue.

There are three ways to deal with an essential resource, the availability of which is becoming restricted. One is to obtain more of it, by whatever means are thought necessary. The second is to make better use of whatever one can get. The third is to find or develop an alternative. If such a substitute is available it will represent the most obvious and practical solution. Looking at alternatives will be the subject of the next chapter.

2

PLANNING FOR OUR FUTURE ENERGY SUPPLY

The modern economy always has been, and still is, based on the energy provided by fossil fuels. Petroleum in particular is essential for all types of transportation, from motor scooters to airplanes. It is also the resource which will, in the coming decades, become increasingly scarce and expensive. As said earlier, there are three ways to deal with this problem:

- Ensure we have an adequate, or at least a minimum, supply of oil for as long as possible.
- Get the most out of the supply we have.
- Identify and develop alternative forms of energy.

It is worthwhile to look at the last item first, because the nature and viability of alternatives will determine how easily and quickly we can wean ourselves of our dependence on oil. Only then will the viability of the first two options become apparent.

CHAPTER FIVE

Are Current Alternatives Adequate?

The first question to ask with regards to the energy issue is: What are the alternatives to petroleum? The broader question beyond that is: Are there alternatives to fossil fuels in general? The answer to both is ultimately the same, but the time scale is different, which is of major importance in practice.

Before an attempt to provide answers, however incomplete, one key concept needs to be clarified; namely, what "sustainable" means. It is often said or implied, particularly in the context of the climate change or "global warming" debate, that the alternative to an economy powered by fossil fuels would be a sustainable one. In other words, that the alternative economy would only use such energy as is derived from ongoing natural processes: sunlight, wind, tides, plant growth, and so on.

Here it is worthwhile to repeat what has already been stated earlier: that the pre-industrial economy was fully sustainable according to the above definition. It used only natural inputs and recycled just about everything; since all goods were made to last as long as possible, there was very little to recycle. The industrial economy, by complete contrast, was, and still is, *an escape from sustainability*, made possible by the expenditure of the energy stored in fossil fuels in earlier geological eras.

A return to sustainability as *existed in the past* is not even conceivable at this time. There is no way the current world population could be fed, clothed, and provided with the bare necessities without large inputs of nonrenewable, meaning fossil, energy. Technologies capable of extracting energy from renewable inputs already exist. They will undoubtedly be improved and refined, and entirely new technologies will in time emerge as well. We can safely say that, ultimately, technological progress will allow us to create a fully sustainable economy. But the task of getting us from here to there must not be underestimated. Even with massive investment in research, testing, production facilities and distribution such an evolution will take decades. We must, therefore, make a clear distinction between two concepts:

- A definitive, *long-term solution*, which we cannot define at this point, and which will take considerable time and effort, to be put in place.

- An *interim solution*, which can be implemented in the short term, makes maximum use of existing technology and infrastructure, and will provide the time needed to plan for the long term. It is this interim solution and related policies that are outlined in this book.

For this purpose, the question to be answered is the following:

Can petroleum, two thirds of which are imported, be replaced by other forms of energy?

There are currently two potential candidates: electricity and bio-fuels.

ELECTRICITY

Before going into some detail about the potential for replacing oil-based energy with electricity, it must be remembered that this is not a new issue. The concept of an all-electric civilization has been promoted and pursued ever since electricity was discovered, without ever being attained. Current proposals leading in that direction must, therefore, be taken with a measure of skepticism. When something long predicted and hoped for does not materialize, there usually are good reasons for it, be they technical or economic, or both.

Electricity as a form of energy is extremely flexible, and its main advantage is that it can be transmitted instantly from the source to the user. Once the grid is in place, all it takes to get power is to flip a switch. The corresponding drawback, however, is that electricity cannot be stored in large amounts. It must be generated according to demand, or used according to the available supply. This requires an active power grid, involving four major components: generation, transmission, storage when unavoidable, and overall control.

Generation

The bulk of electricity used in the USA is currently generated in plants burning fossil fuel: mostly coal, with natural gas and some oil used as well. As far as fossil fuel is concerned, the only potential to substantially increase electricity generation lies in using more coal. This is both economically and technically feasible, but may run into constraints due to concerns about climate change, since coal-burning power plants are a major source of atmospheric CO_2. While the theory of human-induced climate modification through greenhouse gas release is still a working hypothesis and has not been conclusively proven, it nevertheless has considerable support within the scientific community and should induce prudence with respect to large increases in coal consumption. Should the impact of CO_2 on climate come to be fully demonstrated, this avenue will quickly become a dead end. If the use of fossil fuels is to be reduced, there are two alternatives:

— Nuclear power

— "Renewable" sources: wind, solar, hydroelectric and geothermal

The Nuclear Alternative

The generally proposed alternative to coal, for the purpose of electricity generation, is nuclear energy. This alternative, however, is one of those "solutions" that work wonderfully on a small scale and become progressively more impractical as the scale is increased.

For the purposes of this discussion we will leave aside the issue of security of operation, which can be improved through new designs. Also left out will be the NIMBY (Not in My Back Yard) issue, which will fade if the need for energy is dire enough.

Beyond those, any large-scale increase in nuclear generation faces two insurmountable problems: proliferation and radioactive waste.

First proliferation: nuclear power plants both use and produce the very same materials from which nuclear weapons are made. Controlling where these materials are produced and how they are used is already a serious issue today—witness the ongoing U.S. attempts to shut down Iran's and North Korea's nuclear programs. Such attempts to limit the spread of nuclear technology rest on the premise that only "responsible" and well-governed states should be allowed to possess it, a criterion that will always include a good deal of subjectivity.

If nuclear power is generally recognized as a major and legitimate alternative for power generation, then the acquisition of nuclear technology becomes a universal right, since all states can be said to have equal rights to a basic supply of electrical energy. Once this is established, the main argument against the spread of nuclear technology becomes void. The few hundred reactors now operating, concentrated mostly in the developed states of the northern hemisphere, will become thousands, dispersed over the entire globe. With that number of facilities in operation it will become impossible to control either the fuel production pathways or the ultimate uses of the technology aside from power generation. There will inevitably be leaks and lapses of control; fissile materials and weapons technology, from easy-to-make "dirty bombs" to true nuclear devices, will inevitably start circulating underground. As reactors and nuclear fuel facilities have a useful life measured in decades, there will be no way to determine which government is, and will remain, "responsible" and which is, or will become, "rogue." Governments and ruling political parties could shift several times from one category to the other within the lifetime of a nuclear facility with no way to predict their ultimate ambitions and behavior.

The risks of misuse of nuclear technology and materials will rise in direct proportion to the number of nuclear facilities built and to the extent and complexity of fuel processing networks. As the number of facilities and knowledgeable experts increases, universal nuclear proliferation, both official and rogue, will become a *fait accompli*. The likelihood of a nuclear conflict, accident or terrorist-type attack will increase in direct proportion.

We can choose to ignore this problem and let our descendents sort the situation out a generation from now. By then, however, an irreversible spread of nuclear technology might well have taken place. It would be far more prudent to face up to it while we still have the opportunity to make our decision at relative leisure.

The radioactive waste problem runs in parallel to the proliferation issue, with a delay in time of a couple of decades. Nuclear materials, including the components of decommissioned reactors, remain dangerous for extremely long periods, and the quantities to be disposed of safely will vastly increase if nuclear power is accepted as a bona fide mass energy solution. The U.S. government has been searching for a storage solution for decades, even while radioactive waste keeps accumulating at nuclear sites.

The Soviet Union for its part was far less scrupulous in this matter: large amounts of waste were simply buried on land or dumped at sea. Such disposal methods are now recognized as unacceptable. If we have that much trouble dealing with the limited amounts of waste generated today, how will the world deal with quantities that are larger by an order of magnitude? Where will those huge quantities of dangerous waste be stored, how will they be transported to that location, and what authority will supervise the disposal? The question is not simply practical, but is an ethical one as well: who will take the responsibility of imposing that burden on future generations, and on what grounds will such a decision be justified?

For the above reasons we consider the large-scale use of nuclear energy an unacceptable option. It is possible that *fusion*, as opposed to *fission*, will at some time in the future offer a more acceptable alternative. But nuclear fusion, if it ever comes, remains a potential solution for the long term only. We cannot count on it now.

Renewable Sources

Electricity generation from renewable sources includes hydroelectric, geothermal, wind, and solar. Hydroelectric power is renewable, but the locations for the installation of dams are limited, and the best ones are already occupied. Wave and tidal generation offer promise, but are still at an experimental stage, and will not make meaningful contributions for many years.

Geothermal generation most closely resembles generation from fossil fuels, as it uses heat as input, and can deliver energy on demand as well. It is, therefore, a ready replacement, but for the time being its use is limited by the lack of experience and the availability of suitable locations.

Solar thermal generation offers similar potential, with a greater choice of locations, although the best ones are located in desert areas away from population centers. The technology, however, needs to be developed on an industrial scale and the cost of the power that would be produced is still to be determined.

Wind has been used extensively since early times and is a well established technology. The key problem lies in scaling up power generation because of the intermittent character of the wind source. If wind power comprises less than 10 percent of the total power fed into the grid, the variations induced by weather can be compensated by generation from other sources. As the fraction of wind power grows, the weather-induced variability creates larger and larger problems, forcing unpredictable cutbacks in consumption.

The general problem with all renewable generation is of course cost. None of the above methods would be under consideration if their use was not heavily subsidized. The need for subsidies is not only a financial issue. It also suggests that the capital inputs into renewable energy sources are high, which raises the issue of the "energy margin" already discussed with respect to petroleum. If one calculates the total amount of energy invested in manufacturing, installing, running and maintaining a renewable generating facility, how much of an energy surplus does that facility produce over its expected lifetime? Exact numbers are not available, but the cost situation appears to indicate that the "energy margin" of renewable generation is considerably less than that of generation based on fossil fuels. That in turn means that a number of current economic activities, powered by electricity, would become unaffordable, from an energy balance point of view, as more renewable power came on stream. The value of those activities for our overall economic life would then need to be determined together with the changes resulting from their elimination. Only when these impacts are quantified can a large-scale effort to develop renewable generation be undertaken.

Since increased reliance on electricity, especially for transportation, would increase overall power requirements, such evaluation and planning are needed before specific policies are decided upon. It is of no use to add generating capacity if the cost of power rises to levels where significant sectors of the economy are priced out of the market, with the resulting loss of employment. Or, at the very least, such effects should be foreseen and provision be made for them.

Transmission

Electrical power cannot be stored on a large scale in the same way oil or gasoline is stored in tank farms. If a means to similarly store electricity were found, many of our energy problems would disappear, but the search has been fruitless so far. Electric current must be transmitted, more or less instantly, from the point of generation to the point of use. Power is lost along the line, proportionally to the distance the current travels. For this reason, generating plants are usually sited as close as possible to the point or area of use.

With fossil fuel plants such location is nearly always possible. With nuclear plants, somewhat less, due to the need for large quantities of cooling water. By contrast, generation from renewable sources such as wind is tied to the most favorable natural locations, which tend to be away from population centers. In the continental U.S. the most favorable areas for wind power are the high plains and the ridges of the Rockies and the Appalachians, about as far from areas of industrial activity as is possible.

Large-scale use of wind resources will thus require the development of a large network of new power lines. To reduce transmission losses these lines would have to use DC current at very high voltages. Extensive easements and rights-of-way would need to be negotiated, with the attendant legal and environmental problems. Connection to the existing grid would need to be worked out, together with control mechanisms to account for weather variability.

This will represent not only a major technical undertaking. There will be large funding requirements; issues of ownership, rates and profit will need to be settled, as well as priorities in the delivery of energy in case of drops in supply. All the above imply some form of national control over the grid, to be worked out by the owners, operators, and relevant governments.

An additional issue requiring considerable attention is security. Long transmission lines in sparsely inhabited areas are highly vulnerable, both to disruption by natural phenomena such as storms and heavy snows, and to human action. It is far easier and faster to bring down a power line than it is to cut an underground pipeline. If a pipeline is cut, oil-based products can be moved by barge, truck or train. For electricity the transmission line is the only mode available. The more economic activity relies on electrical power, the more such potential disruptions, be they from natural causes, internal sabotage, or enemy actions, must be taken into consideration. Not only measures must be taken to protect the lines, but the grid must be capable of compensating for transmission interruptions.

Storage

As mentioned above, electrical power cannot be stored in large quantities. All known storage devices can deliver continuous power for only a limited time. In addition their energy content per unit of weight is low: a car battery will deliver far less energy than the equivalent weight of gasoline or diesel fuel. Recharging a battery pack takes much longer than filling a fuel tank or feeding coal into a furnace. Thus, while electricity is well suited for static applications where a connection to the grid is possible, fully mobile ones present major challenges in terms of both range and power delivered. This is especially significant in the case of long distance, heavy duty transport, used for the bulk of strategic and essential goods. Electric transportation is possible and even advantageous within areas of high population density, but its potential rapidly falls off with distance. An economy powered by electricity would only be possible with much higher population densities, such as existed in the early twentieth century within cities internally connected by rail and tram lines. Moving back from our current suburban sprawl into such densely packed cities would imply a major cultural change and require a complex new set of policies.

Control

Our current power supply and its utilization is based on a close to optimal compromise between the respective advantages and drawbacks of electricity and direct fossil fuel consumption. The more functions are turned over to electrical power, the more foolproof, redundant, and reliable the power grid must be. We now can use both energy sources interchangeably, as in having a gasoline-powered generator should our home power fail. Our economy is more tolerant of grid breakdowns and blackouts, and we can do without electrical power for limited periods. But as one moves from such a mixed energy delivery system to an all-electric one such breakdowns become far more significant and potentially dangerous, not only in peacetime but even more so in wartime or in situations involving internal disturbances.

The design and building of such a power grid, with its built-in redundancies, fail-safe features, and abilities to match variable power inputs to changing demand will present significant challenges in terms of control technology as well as investment. Aside from the issues connected with controlling the grid, the key challenge is that of distance. Power losses in transmission are proportional to distance, and power lines are expensive to build, maintain and protect. Therefore, the closer our economy comes to being all-electric, the more economic activity will need to be physically relocated. It will also have to be re-programmed in time so as to balance power supply and demand. This issue must not be underestimated as it will have not only an economic impact, but a social one as well, affecting population distribution and density, the relationship between home and workplace, and many aspects of our lives that have been around so long that we no longer question them.

RENEWABLE FUELS

Transportation would be the weakest link in an all-electric economy, and the one area where fossil fuels, particularly those derived from petroleum, offer truly unique advantages. To overcome this difficulty the use of renewable fuels, also termed "bio-fuels," has been suggested. Renewable fuels are understood as those derived, through a variety of processes, from plants, algae, and other living organisms. The fuels currently under consideration to replace oil-based products are biodiesel, derived from vegetable oils, and ethanol, derived from corn in the U.S. and from sugar cane in tropical countries.

One of the major arguments for the development of these fuels is that the feedstock is renewable. This, in theory, would eliminate or reduce the CO_2 emissions associated with fossil fuels and held responsible for global climate change. This is a questionable benefit as significant inputs in terms of fossil energy are needed to produce the original bio-fuel feedstock as well as for processing it into finished fuel. This makes the CO_2 reduction much smaller than originally expected, possibly eliminating it altogether. But the primary issue with renewable fuels is not just there. Even if more efficient production processes are developed, yields are raised and land otherwise unsuitable for agriculture is used, the rate at which these fuels can be produced will, in the end, be limited by the growth rate of the plants or organisms providing the raw material. Renewable fuels thus can provide some relief from a total reliance on petroleum, but they will allow only for a static level of economic activity, not the constant increase hitherto made possible by the use of fossil energy sources. In the final analysis "renewable" is equivalent to "sustainable," and sustainability is, in the current state of technology, a limitation as well as a benefit.

We are left with two conclusions:

First, that there is a fundamental contradiction between the concept of sustainability and the way our economy has operated for the last century and a half. An economy fueled primarily by the burning of fossil fuels, as ours has been and still is, offers the potential of continuous material growth as long as the fossil fuel reserves remain large and accessible. By contrast, a "sustainable" economy is limited in its energy use by the rate at which such energy can be extracted from cyclical natural processes, which proceed at a set rate. The more we rely on "renewables" the faster we will have to abandon the growth paradigm. The increased use of renewable energy *as it is currently conceived* does not provide us with a time window to develop long-term energy solutions. In actual fact *it reduces that window.*

Research into energy alternatives to fossil fuels, and specifically petroleum, should not only continue, but it should be increased and given high priority. As research progresses the various options should be evaluated, tested and rated. Planning to implement the more promising ones should begin. But we should not rush into premature implementation, conflating the interim with the long term.

The solutions currently being contemplated, such as greater reliance on electricity and on "renewable" energy, do in fact require years, if not decades, to be properly evaluated and implemented. There are other technologies that need to be studied and tested as well.

The growing constraints on the availability of petroleum and related products such as natural gas mark the *beginning* of the end of the era of pure quantitative growth. I want to insist on the word "beginning" because those constraints have grown, and will continue to grow, in a gradual manner. There will be times when they will be relaxed, due to lulls in demand or other circumstances. In addition, the restrictions applying to petroleum have not yet been generalized to other fossil fuels, of which the main one is coal. We are not, whatever might be said to the contrary, "running out of fossil energy." We are not even "running out of oil." Nevertheless the warning bell has been rung, and it will continue to ring, louder and louder as time passes. But in the meantime we still have time to study, think and plan.

In order to do this right, we need to stretch out our use of petroleum a little further. So the second conclusion is: How can we secure our petroleum supply for the near future, so as to give ourselves that time?

CHAPTER SIX

Oil and War

The conclusion of the previous chapter was simple: there is currently no viable substitute for oil. While we definitely need to develop alternatives, these will require considerable research, development and investments before they can satisfy a significant portion of our energy needs. So we must be able to continue using oil for the intermediate period.

This means assuring ourselves of a secure and stable supply. Without this our economy will not be able to function, and our standard of living will fall significantly. More importantly for the long term, energy shortages will deprive us of the opportunity to develop long-term solutions.

Currently the U.S. consumes roughly twenty million barrels a day of crude oil. About two thirds of this amount is imported, and one third is produced domestically. We will look at both sources in sequence.

DOMESTIC OIL PRODUCTION

The United States was once the world's top oil producer. During WWII American petroleum resources fueled the bulk of the allied war effort, but over time domestic production has fallen far behind consumption. There is still considerable potential for oil discoveries on U.S. territory, particularly offshore, where many areas are currently closed to oil exploration. There is also considerable potential in the Arctic, the exploitation of which would be facilitated by the current retreat of the ice pack. New drilling and recovery techniques also allow exploiting geological formations previously considered beyond reach.

Potential development, however, must be matched against the depletion of existing currently exploited fields. Total domestic production,
old and new, decreases each year by roughly a quarter million barrels
per day. To keep this production at the current level would require the
discovery of a million barrels-per-day field every three or four years,
which is unlikely to happen. The last such field to be discovered, Prudhoe
Bay in Alaska, was found in 1968. Even if another one was located,
several years would be required to bring it into full production. It is
reasonable to assume, even while pursuing exploration and drilling,
that in all probability new discoveries will only slow the rate of decline.
Progress in recovery techniques, as well as the reopening of depleted
fields due to the higher oil price, will have a similar effect: beneficial,
but insufficient.

Domestic production, therefore, still has considerable potential,
and its development must be pursued. Exploration and drilling must
continue, and in particular the highest priority must be given to the
development of new technologies. Nevertheless, additional sources of
supply are needed, which means imports.

IMPORTS: GOING CRITICAL

Imports have filled the gap between total U.S. demand and domestic
supply. They have provided a satisfactory source of oil until approximately
2000, but that situation is now changing, primarily because world
demand has caught up with global production capacity. This leaves the
market vulnerable to production delays, due to difficulties in bringing
new fields on line or to local insecurity. To these must be added attempts
by oil producing states to manipulate prices.

Insecurity, shortages and manipulation make for a volatile market, which is the opposite of a stable supply. There are two ways to
acquire something that is in short supply, which we need, and which
someone else owns. One is to secure it by force. The other is to create
an arrangement beneficial to both parties. The first one is tempting

to the one who has the power, but it does not work. To find out why, let us look at Iraq.

THE WRONG WAY TO GET OIL

Since the oil market is global and roughly two thirds of world oil reserves are currently in the Middle East, the U.S. energy policy has been focused on ensuring that access to Middle Eastern oil remains free and that the U.S. has a place at the table. Thus, no discussion of energy supply can avoid a parallel analysis of U.S. foreign policy with respect to the region and its conflicts. Among these the war in Iraq has been a stand-out, and can serve as a model for analysis.

The war began in 2003, and has cost thousands of American lives, as well as Iraqi lives in the tens of thousands. The cumulated direct cost to the U.S. will certainly reach a trillion dollars; the cost to Iraq is hard to estimate at this point, but it is enormous. Despite the improvements in security after 2007, the war and the U.S. occupation are still not over, with expenses running at ten billion dollars a month. In addition, the bill for future replacement of worn out or destroyed military equipment is increasing daily.

A number of successive justifications have been provided for the war: first, the elimination of Iraqi weapons of mass destruction; next, the capture of Saddam Hussein and the dismantling of his regime; after that, the creation of a democratic government; then the destruction of Al Qaeda in Iraq; the next goal was to stabilize the Iraqi government; etc. None of these, however, can adequately explain either the length of the conflict or, more importantly, the lack of proportion between the expense incurred by the U.S. and the hoped for results. One is, therefore, driven to ask whether the war is not, in reality, about something much more important, strategic, and valuable than all of the goals given so far for the U.S. effort; namely, oil. It is not possible to be sure without having been privy to the inner debates within the Bush administration, but one can at least examine whether the "oil explanation" makes sense.

Here are the known facts: any U.S. forces in Iraq are within a couple of days' striking distance of a least half of the world's oil reserves, or two thirds if one includes Iran and the Caspian states. The United States has visibly been planning for an indefinite stay: witness the fortress-like U.S. embassy, the largest in the world, built at considerable cost in the "Green Zone" of Baghdad, as well as the numerous other bases the U.S. forces have built or expanded. The U.S. Navy has extensive facilities in the region, and is capable of supporting land forces for long periods. Without taking into account Iraq's strategic location and own considerable oil reserves it is hard to see why such an immense effort is being undertaken for the sake of reforming a relatively small country with a population of 26 million.

The "oil theory" thus fits the facts very well. Before we go further it is important to note that this statement, per se, is not meant either as criticism of the U.S. administration, or as an expression of support. The key questions are, first, whether a policy of occupying Iraq in order to ensure U.S. access to, and control of, the vast oil reserves of the Middle East is a workable one. Second, whether an alternative policy, or set of policies, could better serve the goal of providing the U.S. with a secure oil supply.

While considering the first question, it must be noted that only a tenth, at most, of the adversaries U.S. forces has been facing in Iraq were part of, or connected with, Al Qaeda or similar jihadist and/or terrorist organizations. The truth is that Al Qaeda established itself in Iraq only *after* the U.S. invasion, taking advantage of the already existing insurrection. The conflict has always been primarily with native Iraqi insurgents. The longer U.S. occupation lasts, the easier it will be for the locals, both from Iraq proper and from the wider region around it, to believe that the U.S. is indeed there for the oil—in other words, that it seeks to make Iraq into a kind of colony just like the British tried to do after WWI, and for some of the same reasons: control of the oil fields. Once such a belief is firmly anchored within the population, insurgency is likely to become permanent as well. We would then end up fighting a typical colonial war, and if past history is any guide we are not likely to win it.

The oil industry is one of the hardest to control and defend by militarily means. The oil fields themselves are dispersed over vast swathes of territory so that the protection of the wells requires large forces, up to date intelligence, and a number of strategically located bases. Pipelines are long and extremely vulnerable. The product is flammable, so that any successful attack will cause considerable damage resulting in long repairs. The physical extent of the areas needing protection increases the risks associated with ambushes, sabotage and other forms of guerilla warfare.

The forces needed for such attacks on the oil infrastructure need not be large; in fact, keeping them small and mobile is more effective. Their support, either by the indigenous population or by a foreign entity, will be relatively easy and inexpensive; by contrast the maintenance of the security forces, obliged to protect all installations around the clock, will require far greater manpower and resources.

To verify the above, one need only look at the present situation in the Niger delta, where the bulk of Nigeria's onshore oil facilities are located. A few bands of insurgents and local bandits have managed, with little effort and expenditure, to shut down between one quarter and one third of the country's oil production, through highly targeted strikes. And yet, they are only fighting *their own* national government. The insurgency would likely be far more intense were it directed against a foreign occupier.

The other example of the costs and damage incurred is provided by the U.S.'s own results in Iraq. Roughly 15 percent of the population has been displaced, either internally or by moving abroad. Unemployment and under-employment are rampant. The professional class has mostly emigrated. The infrastructure is still in shambles despite a five-year rebuilding effort and plenty of available funds. The oil production itself is barely back where it was under Saddam Hussein. Oil companies have stayed away from the main oil fields due to the poor security situation, and no large oil development projects have been initiated, save for a few minor ones in the Kurdish north.

Over half a decade of military occupation has been a major drain on U.S. resources and the region is more unstable now than it was before our intervention. If we want democracy in the Middle East, we are going about it the wrong way because military occupation is the very antithesis of self-determination. If we are after oil, we are going about it the wrong way as well, because the oil fields cannot be taken and held by military means. There are good reasons to consider alternative approaches. These will be taken up in the next chapter.

CHAPTER SEVEN

Better Ways to Obtain Fuel

If we accept that a military occupation of portions of the Middle East will guarantee neither our supply of oil nor stability in the region, another strategy must be worked out. This strategy must have a realistic chance of achieving the following goals:

- Provide a secure fuel supply to the United States—including U.S. armed forces.

- Relieve the pressure on global petroleum supplies so as to avoid or mitigate the economic consequences of a continuous rise in oil prices, and specifically oil spikes such as occurred in 2007–2008. As the U.S. imports two thirds of the oil it uses our economy is very vulnerable in this respect.

- Provide the U.S., in case of urgent need, with a guaranteed oil supply.

- Allow for an orderly drawdown of U.S. forces in the Middle East while still maintaining stability in the region.

The recommended strategy is as follows:

1. Secure Fuel Supply

Using the word "fuel" instead of "oil" is deliberate. We have stated earlier that there are no ready alternatives to petroleum as a source of energy. There is, however, an alternative in terms of fuel supply. This option is not the ultimate solution, since it relies on fossil hydrocarbons for input. But as a bridge solution it has considerable merit.

Transportation fuel can be made, and currently is being made, from coal or natural gas through what is known as the Fischer-Tropsch (FT) process. This involves turning the coal into "synthesis gas," which is then run over specially designed catalysts, producing what can be termed "synthetic petroleum." This liquid can then be subjected to a treatment similar to what oil undergoes in a refinery, producing a range of hydrocarbons (fuels, lubricants, and other products) similar to those currently derived from petroleum. Because the process can be accurately controlled these products are purer and more homogenous than those actually made from oil. Another advantage is that the end-product can be optimized for its intended use through control of the chemical reactions.

The process was developed in the 1920s in Germany and was used to great effect in WWII, allowing the German armed forces to continue military operations long after their sources of natural petroleum had been cut off or destroyed. It was later picked up and perfected in South Africa, which feared an international oil embargo due to its racial policies. Synthetic oil installations using coal as raw material still produce a substantial portion of the fuel South Africa uses. Sasol, the corporation that holds the relevant know-how and patents, is currently designing or building synthetic fuel plants worldwide, using coal or natural gas as input. Other companies, such as Shell, have developed their own expertise, patents and processes. Fischer-Tropsch facilities using organic feedstock (such as wood or agricultural waste) are being tested and developed as well.

The FT process is a mature one, has been tested on an industrial scale, and is economically competitive with oil above $60/barrel. FT aviation fuel has been certified for commercial use by the U.S. Federal Aviation Administration. The advantages of synthetic fuel have not escaped the U.S. Air Force, which consumes nearly three quarters of all the fuel used by our armed forces. Concerned, like other branches of the military, that the U.S. supply of imported oil might someday be cut off, the Air Force is currently running extensive flight tests to accurately verify the performance of synthetic jet fuel in a number of key areas: energy content, ability to mix with oil-based products, shelf life, and other properties. All tests, including supersonic flight, have yielded excellent results so far.

The United States has the world's most abundant coal reserves, which are both highly accessible and widely distributed throughout U.S. territory. There is no major issue in raising coal production over time, and since there is no shortage of the raw material its price would not rise substantially. There is, therefore, no obstacle to the initiation of a synthetic fuel program, answering the wishes of the military as well as the demands of citizens squeezed by rising fuel prices.

Such a program would be far less expensive than the ongoing wars. A synthetic fuel plant producing 80,000 barrels of high-grade, low-emission diesel fuel per day would cost initially around five billion dollars, with the cost dropping and output rising as more plants come on line and the technology is incrementally improved. If one uses the current $10 billion per month figure for the war expenses in Iraq, the U.S. could build 25 or more such plants for the cost of one year of war. These plants would produce a total of 2 million barrels of ready-to-use fuel per day, or the equivalent of all of Iraq's current oil production. Both the production facilities and the source of raw material would be located safely in U.S. territory.

There is of course no way to build such a number of installations within a year, so the above comparison is for cost purposes only. Nevertheless, had such a policy been adopted at the time of the initial Iraq invasion, we would today be well on the way to a real improvement in strategic security, with a few plants in production, several others under construction, technology advancing and costs being reduced. The world would have been put on notice that the U.S. is serious about energy security and is now acting to reduce its dependence on oil imports. The effect would be lower demand for petroleum, a lower price, less pressure on oil fields and an extension of their useful life. Other countries with available coal reserves would then consider following the U.S.'s lead, to the benefit of their citizens and of the world in general.

One additional advantage of the FT process is its suitability to produce diesel, the main fuel for heavy transportation, as well as kerosene or jet fuel. Increased supplies of diesel, in particular, would lower its cost and favor the replacement of gasoline engines by diesel-powered ones. Since the latter are 20 to 30 percent more efficient than the former, this in itself would lead to a significant reduction in oil demand as an add-on effect.

Current FT technology also allows to quickly ramp fuel production up or down. As the raw synthesis gas produced from coal is combustible, it can also be used to run turbines or boilers. A synthetic fuel plant can then be coupled with an electrical generating facility, increasing electricity production when demand for current is high and ramping up fuel output when such demand is low.

Such an arrangement is complementary to producing electricity through wind power. If the wind blows, the combined FT and power generation plant will produce fuel, which is stored and fed into the distribution infrastructure. If power demand increases, the plant switches over to power generation. This solves the "back-up" problem, or the need to have fossil-fuel-fired facilities to provide back-up power for wind farms. Such single-purpose facilities would sit idle whenever the wind blows, thereby increasing grid cost. By contrast a combined FT fuel and power generation plant would never be idle.

The FT process is not a panacea, however. While it delivers cost reductions and strategic security, it has an environmental impact in the form of more coal mining and the production of large quantities of carbon dioxide, a greenhouse gas. The process does, however, allow for easy separation of the excess CO_2 from the synthesis gas used to make fuel. This leaves room for adding a carbon sequestration unit to each facility, if and when required.

Synthetic fuel production offers a ready-to-use and economically competitive alternative to oil imports. It will not, nor does it have to, replace all twelve million barrels of crude now entering the country from abroad each day. All it needs to provide is an ongoing "strategic reserve" as well as a damper on prices. It will also demonstrate that, to a degree, "we can do without oil," a mental step which must be taken if we are to progress in time toward an economy no longer dependent on fossil energy.

Once a synthetic fuel program is implemented as a matter of national policy, the psychological impact would be immediate. The physical one would be delayed by the amount of time it would take to bring a significant number of installations on line. This would normally take three to five years, even though Americans have been known to speed things up when really needed, as in WWII.

There will still be, in the short term, a need for a secure source of emergency oil in case of a crisis. This is possible as well, contingent on an adjustment in our foreign policy.

2. Emergency Oil Supply

So far our primary focus in terms of oil supply has been on the Middle East. If one considers the potential of synthetic fuel production within the United States, that region's reserves, while still important, become less critical. Even with synthetic fuel, there is no way, in the short term, for the United States to achieve full self-sufficiency in petroleum products. Since full control over foreign supply is not attainable, could we at least obtain a guarantee that such supply will not be interrupted? In other words, is some kind of long-term, preferential arrangement with one or more suppliers possible?

As the previous analysis of the Iraq conflict indicates, the Middle East is in all likelihood *not* the place to seek such a long-term arrangement. The very concentration of reserves makes the oil market in that region highly competitive, and there is no reason why the states controlling the oil fields would not demand the best possible price, and sell to the highest bidder. On the one hand, oil is their greatest source of wealth and influence, and they are unlikely to squander this unique resource by providing overly generous terms of sale. On the other, as we have shown earlier in respect to Iraq, it would be extremely difficult as well as prohibitively expensive for a single power to control the region through military force. To achieve this, such a power would have to face fierce and long-lasting local resistance as well as competition from other oil-consuming states. These difficulties would grow in scope and intensity as the availability of petroleum gradually decreases.

Thus, while it is of interest to all that the Middle East remains relatively stable, that region will, by the same token, remain an open and competitive market, unlikely to favor special arrangements. Another venue would better suit our needs, and in fact there is one with considerable potential, where a preferential arrangement could work to the mutual benefit of all parties concerned.

There is one large, relatively unexplored region, where considerable oil resources already are known to exist, and where substantial new fields could still be discovered. This is the northern Eurasian landmass, which is under Russian control. Whereas in Cold War days Russia was the U.S.'s principal enemy, the dissolution of the Soviet Union, coupled with the end of communist power, has now removed most points of contention. It is true that after a period of close cooperation in the 1990s the current decade has seen a renewal of hostility. But most of this disagreement has been artificially created for apparently political reasons, particularly the ill-advised U.S. policy of military expansion into Eastern Europe and Central Asia. In actual fact, the USA and Russia have many more common interests than diverging ones, as well as no particular area where their national interests truly require them to compete. In the specific case of oil production and supply, their needs are in fact, highly complementary and can form the basis of a mutually beneficial relationship.

Russia can, at this time, produce enough oil for both its internal needs and for export. The same applies for natural gas, the largest global reserves of which lie within Russian territory. Russia does, however, have two significant problems.

The first is essentially a financial one. On the one hand, energy exports provide the Russian state with a large percentage of its budget revenues. Oil revenues have, in this decade, allowed the government to pay off outstanding debts and to accumulate a significant budget reserve. However, the oil fields currently accounting for the bulk of Russia's energy production were initially developed during the Soviet area. Most by now have reached maturity, with some already entering decline. Untapped fields do exist, but their development will require considerable investment, which would reduce the portion of oil revenues that go to the government. Putting it in simple terms, Russia cannot have its energy cake and eat it too. Oil revenue can be allocated to production investments or to the state budget, but there is not enough to fully satisfy the demands of both.

The other problem is one of technology, and this compounds the capital shortage. Russian oil exploration and extraction technology is mostly of Soviet vintage. While substantial improvements to existing fields have been made through subcontracting work to Western oil service firms, this source of up-to-date know-how is not sufficient for entirely new, large-scale projects. Only the international oil majors such as Exxon or Total have the necessary technical expertise and financial resources.

Herein is a stalemate: the "Big Oil" majors want production-sharing contracts before investing; the Russian government wants *full* control of energy resources as well as of how they are to be allocated. As a result the majors do not have access to the huge present and future Russian energy reserves. And the Russian present "full control" policy keeps it short of development capital and oil field technology. This stalemate benefits no one, while both Russia and the U.S. are losing valuable time as well as access to potentially huge resources and the associated income.

The deadlock can be broken through direct government-to-government negotiation, with the participation of production entities on both sides. Through a bilateral agreement Russia would get outside capital and cutting edge technology; the U.S. in exchange could obtain access to a guaranteed amount of Russian oil and/or gas. The additional oil and gas would be routed through the oil majors, which would thereby obtain the equivalent of the long-term production contracts they seek. The U.S. would acquire a secure source of energy and Russia receive a long-term guarantee of both revenue and energy development.

Such an arrangement could have huge additional benefits in terms of scientific development, technical cooperation and commerce. Russia is currently in the early stages of fully rebuilding its infrastructure, which has badly decayed since Soviet days. Such infrastructure renewal will have to involve much more efficient use of energy resources than is now the case, as the old Soviet methods were extremely wasteful. The U.S. faces a similar challenge in terms of cutting down its own energy consumption. As both countries have highly developed scientific and engineering establishments, there is a large potential for technical cooperation and commercial ventures. The relationship could, in the end, prove highly profitable.

There is one important caveat, however. The complementary situation between U.S. and Russian needs in the energy area will not last forever. If the current, and largely artificial, hostility between the two countries persists, the window of opportunity for cooperation will fade away and ultimately close. Russia will eventually acquire or develop the technology it needs; the U.S. will look elsewhere for supply, and a great opportunity will have been lost.

3. Impact on the Global Energy Situation

What will be the impact of the two policies recommended above? If a decisive move is made toward their implementation, both would have tangible effects within the time span of a single U.S. administration. Within four years a couple of major projects, as well as a number of smaller ones, would be producing their initial barrels of Russian oil, and the first synthetic fuel plants in the U.S. could be coming on stream. This will work toward stabilizing the price of petroleum and to eliminate the inordinate speculation based on short-term price spikes.

The psychological effect would of course be nearly immediate. Once the implementation of the above policies is on its way, oil producers, and particularly those within OPEC, will know that their current stranglehold on the petroleum market will inevitably be loosened. They will understand that the U.S. in particular would not only be receiving specified amounts of Russian-origin oil within a fixed timeline, but also would, within a similar timeline, be initiating synthetic fuel production for its own needs. This will open a window for global negotiations aimed at stabilizing the market by balancing supply and demand around an acceptable price level.

It is to be noted that "the markets" by themselves will not achieve the above results. Over-reliance on the market is in fact what has produced the current situation with its dead-locks and stalemates. Private enterprise will play a major role, but decisive state action will be required as well. Such cooperation and coordination between the public and private sectors will be a general and common characteristic of all policies and measures taken to deal with the requirements of the energy situation.

4. Dealing with Iraq (and Other Related Conflicts)

The policies outlined above will deliver multiple benefits for a relatively low cost:

- A significant increase in the reliability of the U.S. oil/fuel supply, both in terms of petroleum and of a compatible substitute in the form of infrastructure-compatible synthetic fuel.

- Increased leverage by the U.S. on the world oil price.

- Increased oil/fuel production in both the United States and in Russia, providing a cushion against "oil shocks."

- Cooperation between the U.S. and Russia in the field of energy, opening the possibility of cooperation in many other areas, from diplomacy to research.

- Increase in U.S. business and economic activity, with substantial investment and job creation.

Finally, the development of additional oil and fuel resources will relieve the United States of the need to control Middle Eastern oil fields. This will allow the United States to shift its focus from military intervention and occupation to simply fostering and maintaining stability in the region. In practical terms the shift in energy policies will allow to wind down U.S. military engagement as well as the conflicts we are currently engaged in, and to withdraw the bulk of U.S. forces.

This will be the subject of the next chapter.

CHAPTER EIGHT

Stabilizing the Middle East

In the current global energy situation, Middle East means oil, and oil means Middle East. The policies outlined in the previous chapter will create a more stable oil supply situation for the United States, but they will not reduce the size and global importance of Middle Eastern oil reserves, which remain a key energy resource for the entire world. The key condition for these resources to remain accessible is stability in the region and the absence of open conflicts.

In order for such a situation to be created and maintained, the following principles must be accepted as a foundation of policy:

- It is not practically feasible to secure foreign oil reserves through military force.

- The national sovereignty of existing states must be respected.

- The maintenance of stability will require international cooperation.

- That stability in the Middle East, as in any other region, requires that the basic needs and interests of the populations concerned be understood and respected. This is also the fundamental pre-condition of any successful fight against terrorism, as will be explained in a later chapter.

Long-term stability cannot be created through extended military occupation. As the numerous "wars of liberation" of the 20th century have shown, such occupation will inevitably arouse general resistance, be in the form of civil unrest or armed insurgency. Regional stability will, therefore, demand the winding down of the U.S. military presence in Iraq and other conflict areas.

We will briefly review the elements of a drawdown of U.S. forces, and of the issues such a move will involve. Iraq, where our forces are currently the most numerous, and where the issues created by this occupation have been clearly demonstrated, can be used as a model, with the key elements of such a withdrawal being reviewed below.

1. Withdrawal Schedule

The U.S. currently has a withdrawal agreement with Iraq stipulating that American troops will leave the country by 2012. What is needed now is a hard and public schedule for the operation.

Moving an armed force of some 140,000 soldiers, which has been in place for over five years or more, together with its armaments, vehicles, stores and services, is a major logistical undertaking. Regardless of the numerous political statements made, there is no such thing as an "immediate" withdrawal ensuring that the troops will be home by some nearby date. The move must be carefully planned and executed, particularly if the situation on the ground is still fluid. From there it follows that the schedule and timelines must be firm, or else disorder will set in.

In order to be both orderly and credible such a withdrawal cannot be conditional on unspecified future events. The dates of the various steps, and particularly the end date, must be determined in advance and made official. As explained below, this can be an advantage rather than a drawback.

2. Stability in the Country

The most important issue about stability in Iraq is that, if conceived as a unitary rather than federal state, it will be inherently unstable. Iraq as it is known today originated as a British colony-to-be, carved out of the defeated Ottoman Empire in 1918. After a bloody rebellion against colonial rule, the British made the territory nominally independent while retaining major influence as well as access to the oil resources. During this time of theoretical independence, Iraq had more than fifty ministerial cabinets, which gives an idea of the difficulty of governance. In 1958 the monarchy set up by the British was overthrown in a military coup, within which Saddam Hussein had a significant role. Within a couple of years he took over and until 2003 ruled Iraq with the iron hand of a dictator.

Iraq, therefore, has no history or experience as an autonomous, self-governing state, much less as a nation. The population is divided among three main ethnic and cultural communities: the Shia, Sunni and Kurds, with divergent and potentially antagonistic aspirations and interests. Attempting to make Iraq into a unitary nation, especially given that most of the professional elites have emigrated, would be a monumental task, which would require generations to accomplish. The only workable solution for the country is a federal system.

Each community, at this time, has its own militia or militias, which are essentially capable of ensuring basic security in their respective regions. The militias need to be regularized and turned into a form of regional guard, each responsible for its own ethnic territory. A federal zone similar to what exists in other countries can be established around Baghdad. Each region must be given control of its own resources, which is essential for their being economically viable. The Sunni region, where known oil resources are, at this time, limited, can be assisted by an international fund supported by the Gulf States and Saudi Arabia until it has developed its own resources and/or economy.

If all power and resources are given to a central government they will become and remain, as they are now, an object of contention. Each community will try to pull as much of the blanket its way and all three will continue to fight over their respective shares of the spoils. If, on the other hand, control of resources goes with the territory each community occupies, each ethnic group will have a stake in the federal arrangement and will concentrate on exploiting that stake rather than trying to increase it at the others' expense.

There are, of course, disputed areas and conflicting claims, which cannot be completely avoided. No formula will be perfect. If the U.S. is to be rid of the Iraqi problem and the considerable burdens that go with it, it should work out the best attainable arrangement under the current circumstances and do what it can to implement it in the time that is remaining. The U.S. still is, right now, the dominant power in Iraq. It should use this leverage to set up a workable federal system, and then leave.

It has been mentioned earlier that making the withdrawal plan and timeline public from the start, can be turned to an advantage. This connects to the issue of leadership. It is a basic law of history that military conquest and occupation are incompatible with local leadership. No popular leader worth his or her salt will collaborate with a foreign occupier. Such a leader will fight the invader to the death or to victory, as many true leaders, from Simon Bolivar to Kemal Ataturk and beyond, have done in modern history. As long as the U.S. occupies Iraq with an armed force, no genuine Iraqi leader will emerge, except possibly within the insurgency. Announcing our departure well in advance will allow the real local leadership to emerge and take over. These leaders may not be those who run the current government. But they will be effective and capable of setting up a functional government in their respective regions, which is what is most needed for the country to enjoy any form of stability.

Our withdrawal need not be total. A force strong enough to act as arbiter, and to discourage foreign meddling can be left in place as long as the local population tolerates it. U.S. Navy facilities in the Gulf should be maintained, as should mutual help and cooperation agreements with governments there. Finally, some form of multi-lateral understanding and arrangement concerning the stability of Iraq should be worked out as the U.S. withdrawal proceeds, so as to provide a frame of reference for the future of the region and a guide for action. This is particularly important insofar as the entire Middle East is concerned, and is addressed in the following section.

3. Stability in the Region

Whatever arrangements the U.S. can make to ensure its energy supply in the short term, the Middle East will remain a strategic region for the foreseeable future as it is the world's main provider of oil. Relative stability and calm in the region are in the world's interest and it is the responsibility of outside powers, of which the U.S. is one, to contribute to this stability.

If one examines the recent history of the region, it does not come across as particularly unstable or characterized by constant upheavals and revolutions. State borders, although determined arbitrarily in a number of cases, have remained stable, and a relative balance of power between the various states generally prevails. Leaving aside the Arab-Israeli conflict, the main threats to peace in the region have generally come from the attempts of one state or power to exert a preponderant influence, thereby breaking the existing equilibrium. This power can be external to the region, as the Soviet Union was before its dissolution, and as the U.S. is today; or internal to it, as in the case of Nasser of Egypt and Saddam Hussein of Iraq. Looking at the Middle East through realistic lenses, one sees little evidence that the region needs to be "fixed." What one needs to watch out for is the rise or intrusion of potential "disturbers of the peace" seeking to enlarge their share of the pie, or for outside powers attempting to exercise undue influence. During the Cold War, U.S. influence provided the counter to Soviet attempts to dominate the region. Instead of stepping back, and possibly reaching an understanding with the Russians the U.S. government has, from 2001 on, opted for a policy of intervention and armed occupation, which is the least likely to provide long-term stability.

Nasser and Saddam are gone from the scene, and the Soviet Union is no more. If the U.S. terminates the Iraq venture and goes back to a background role, only one threat to the regional power balance would remain: Iran. This has, in the past, provided another rationale for continued U.S. presence in Iraq: that, should U.S. forces leave the southern, or Shiite part of Iraq, where the largest oil reserves are located, would either join up with Iran or fall under Iranian influence.

This last argument is not entirely credible. For the last five centuries the Iraqi Shiites have been occupied and/or ruled in turn by the Ottoman Turks, the British, the Sunni minority under Saddam, and finally by the Americans. It is unlikely that, given the possibility of autonomy within a federal Iraq, they would instead choose to be ruled by Iran. It must be remembered that the Iranians are Persians, while the Iraqi Shiites are Arab. Language, culture, customs, and history are separate and different. Once the Shiites are in control of their oil fields, they are more likely to fight to keep them than to turn them over to the Iranian government.

Nevertheless Iran is a populous country with an authoritarian government and powerful military and paramilitary forces. They are currently engaged in a nuclear development program, which they have not abandoned despite heavy U.S. pressure. Nor are they likely to, which leave military action by the U.S. as a last-ditch option to prevent a nuclear Iran. But even limited hostilities in the area would be the worst possible thing from a global oil supply point of view, and the Iranians know this. So for the last several years we have had a stalemate.

It is here that a policy of collaboration with Russia, already recommended above as a means to secure a better energy supply, can have great potential. The last thing Russia wants is a nuclear Iran, as its territory already is within range of Iranian missiles. Nor does it want an aggressive rogue state close to its southern border. At the same time Russia has a host of internal issues to deal with before it could attempt to exert major influence in the Middle East as it did in Soviet days. Therefore, the possibility for collaboration with the U.S. to resolve the Iranian issue and further stabilize the Middle East is good—as evidenced by Russia's cooperative stance in both the 1991 Gulf War and the 2002 U.S. operations in Afghanistan. In the case of Iran, it is doubtful that its government could face up to Russian pressure from the north *and* U.S. pressure from the south and west.

But outside pressure is not the only means to keep Iran contained. Federalizing Iraq, instead of the attempt to build a unitary state, would provide additional, and possibly much greater, leverage. The population of Iran itself is only 60 percent Persian. The remaining 40 percent are non-Persian minorities. If the minorities of Iraq become autonomous within a federal structure, the minorities of Iran will be inspired to make similar demands, raising issues that would distract Iran from the extra-territorial ambitions it can ill afford anyway.

A Middle Eastern policy aimed at ethnic and cultural self-determination would fit far better with the U.S. democratic tradition than the current one of military intervention. The current policy is a constant reminder to local populations of conquerors and occupiers such as the Ottoman Turks, the British, and the Soviets. A shift toward a more hands-off approach would go a long way toward repairing the U.S. image in the region. How beneficial such a policy can be is demonstrated in the one part of Iraq that is secure and prosperous—the Kurdish territories. The Kurdish population is genuinely grateful to the U.S. for the protection afforded against Saddam Hussein's exactions in the 1990s, and has been generally supportive during the Iraq conflict. There is no doubt that such support could become much wider if U.S. policy recognized the genuine aspirations of other ethnic entities throughout the region.

As far as global oil supply and Middle East stability go, there are better alternatives for the U.S. than a prolonged, and extremely expensive, military presence in Iraq, or for that matter anywhere else in the region. This requires that a few words be said about the "other conflicts" we are currently engaged in: Afghanistan, Somalia, and to some degree Pakistan.

4. Other Conflicts

Besides Iraq the U.S. is actively involved in several other conflicts in the Middle East: Afghanistan, Pakistan and Somalia. These need to be terminated as well if the region is to regain stability. These conflicts are somewhat peripheral to the oil issue, but they have considerable impact on the overall situation in the region as they are socially divisive and tend to feed extremism.

These interventions by U.S. armed forces are all waged under the umbrella of the "war on terror" concept, which has been applied to Iraq as well. This concept assumes that the U.S. is under threat from a massive radical Islamist conspiracy or movement, aimed at destroying Western civilization in general, and the U.S. in particular. The threat from this alleged "Islamic army" is assumed to be so great that resisting it is an overriding national security interest, requiring military intervention in any place where the threat is perceived to exist. Such U.S. response is seen as the carrying out of a "war of generations" requiring that determined efforts, mostly of a military nature, against this assumed enemy be maintained over decades. The "war on terror" concept lumps all Islamist movements and tendencies (be they radical, moderate, or purely religious and philosophical) under a single broad label, which also covers at the far end the most murderous and extreme terrorist groups and entities. Such a generic definition leaves a lot of room for interpretation and makes it easy to intervene unilaterally in any area where Islam is practiced, with the Middle East as primary target.

There are several flaws in the war on terror concept, which make it unsuited as a foundation for any broad and coherent strategy:

The first flaw, which has seldom or never been noted, is that this alleged Islamic movement or conspiracy was only discovered after the September 2001 attack on the World Trade Center. Whereas other militant and aggressive ideologies such as Communism underwent a long and public development before becoming a threat to the established order, there is no such past history for the radical Islamic threat assumed by the "war on terror." Terrorism is of course a serious threat, but to go from the existence of a number of radical cells and groups to the presumption of a transnational "army" is an unfounded stretch of the imagination. No hard evidence of such an "army" exists. There certainly are radical groups connected by shadowy networks and speaking through evanescent Web sites. But there are no visible *transnational* organizations, as opposed to local ones, promoting world-wide Islamic radicalism. In fact the mass rallies, paramilitary formations, and disciplined political parties that characterize international movements were much more in evidence in the earlier period of secular pan-Arabism than they are today.

If one excludes the movements and forces involved in the Arab-Israeli conflict, which is a well circumscribed territorial dispute, the material evidence for a powerful and militant worldwide movement is simply not there.

The second questionable assumption is that this alleged pan-Islamist movement is growing and receiving increasing support among Muslim populations. Again, there is here a blanket amalgamation of every kind of Islamism, which contradicts all studies and surveys carried out recently on this subject. These show that support for Islamism in general, by this meaning the integration of mainstream Islam in political, social, and economic structures, is indeed slowly growing within the Muslim world. However the support for radical militant policies and tactics has been dropping precipitously. This reduction in support is of such magnitude that Al Qaeda itself has bitterly complained about the lack of sympathy for its aims among the Muslim masses.

The third misconception is what appears to be fundamental misunderstanding of the nature of terrorism and of the organizations that practice it. The war on terror mindset makes Al Qaeda and other similar outfits into the vanguard of a much broader and deeper wave, a kind of elite force to be followed by a much larger army. In fact, neither Al Qaeda nor any terrorist group is, or has ever been, anything of the sort. As discussed in the later chapter on the subject of national security, such groups represent an ideological dead end rather than a beginning. No terrorist group in history has ever achieved a significant political goal, and Al Qaeda conforms to this general pattern.

A broad and poorly defined concept such as the war on terror can be politically useful, but it also makes for disjointed and confusing strategy. This is evident in the other Middle-eastern conflicts the U.S. is currently engaged in. A brief analysis of these conflicts and recommended strategies follow.

AFGHANISTAN

The 2002 invasion of Afghanistan was entirely justified and brilliantly executed. The follow-up has suffered from the same flaws as the later operations in Iraq, primarily the attempt to treat as a nation what for centuries had been mostly a confederation of tribal and ethnic entities. Immediately after the initial success against the Taliban, the U.S. repudiated the ethnic leaders and militias that had gained the victory on the ground and settled for a policy of long term occupation supported by a manufactured central government. Military occupation and continued operations against the Taliban were given absolute priority over economic development, leading to the growth of an economy based in great part on illegal drugs and the attending corruption. Both policies squandered much of the good will gained by the expulsion of the universally hated Taliban government, resulting in a military stalemate absorbing ever increasing amounts of resources. Neither a realistic definition of victory nor a coherent exit strategy are anywhere in sight.

PAKISTAN

The policy errors that affect Afghanistan have their parallel in Pakistan, where the U.S. has not directly intervened in an overt manner, but where U.S. pressure resulted in a low-level civil war, paradoxically strengthening a badly mauled Al Qaeda and associated radicals. The insistence that the Pakistani government send its troops into the fiercely independent frontier regions resulted in a small "war of independence" from which Al Qaeda received both direct and indirect support, allowing for a partial rebuilding of its organization. It also emboldened the radical Islamists throughout Pakistan despite their overall lack of popularity, and weakened both the government and the armed forces. With a new government now in place the U.S. is basically back to square one.

The results of our policies to date have been almost entirely negative: the previous President of Pakistan, a long-term U.S. ally, has been hounded out of power; the armed forces have taken a neutral stance while the new government is much more likely to compromise with the insurgents as long as they do not threaten it directly; the population has become mostly hostile to the U.S.; chaos reigns in the tribal areas adjacent to Afghanistan, allowing Al Qaeda to survive and possibly to regain strength. As in other conflict areas, over-emphasis on conventional military force and the occupation of territory has produced a stalemate with no clear strategy to conclusively end the conflict.

SOMALIA

Afghanistan was won, and subsequently lost, for the most part. Iraq could have been won but never was. Somalia has been a loser from the start. Here again the U.S. over-reacted to the fact that an Islamist movement was in the process of taking over the territory and rushed to the "mailed fist" approach. Since no U.S. forces were available, a military invasion by Ethiopian troops was arranged and supported. It was probably also paid for out of U.S. funds, since Ethiopia has no spare resources for such military adventures. The results were the same as elsewhere: substantial killing and destruction, popular hostility, a developing humanitarian disaster, a further breakdown in law and order, and no end in sight. The remains of the Islamist forces defeated by the Ethiopians have been radicalized and a hard-line faction now holds considerable territory even as the discouraged Ethiopians have moved their forces out.

In all three areas mentioned above, the same mistakes are evident as were made in Iraq: a disregard of legitimate ethnic, cultural, and religious aspirations, going directly against America's professed desire to promote democratic trends and structures; a misunderstanding of the Islamist phenomenon and the bundling of the many strands of

Islamism into a single package labeled "Terrorism"; an over-reliance on military instruments of policy; a systematic disregard for the tremendous benefits of economic development; and finally, a refusal to face up to the consequences of our policies in terms of collateral damage: not only fatalities and physical destruction, but also huge numbers of refugees and displaced persons, the collapse or decay of infrastructures, the massive economic dislocation and unemployment, the emigration of skilled professionals, the growth of corruption, and so on. While the soldiers we have sent abroad have done a more than honorable job under very difficult circumstances, the same can hardly be said of the government that sent them there.

The solution for all of the above conflicts is essentially the same as has already been proposed for Iraq: an orderly and well planned withdrawal of the U.S. forces and/or their proxies, leaving behind a structure based on the recognition of ethnic and cultural realities rather than on a theoretical notion of the unitary state. The issues specific to aggressively militant networks and terrorist groups need to be attended to, and this will be addressed in a later chapter. But what needs to be kept in the forefront is the unchanging historical truth and lesson; namely, that the problems generated by the military occupation of a foreign land always tend to be, for the invading power, far greater, much harder to handle, and much longer lasting than those stemming from a withdrawal. It is always wiser to cut one's losses than to keep throwing good money after bad.

ADJUSTING OUR POLICY

Why are these conflicts important? There are several reasons:

First is the issue of regional political stability. As said and repeated previously, the Middle East is inseparable from oil, and oil from the Middle East. Both the Middle Eastern states and their energy customers need it: the states need it in order to use their wealth for the benefit of their populations, the oil buyers so as to have access to a secure energy supply.

Neither war nor military occupation will provide stability. On the contrary, they create a power vacuum, as is evident in all four theatres where the U.S. has been militarily active. Such a vacuum is not only the opposite of stable order, but it invites and empowers the very radical elements that our policy aims to intimidate and destroy.

Second, because the rational exploitation of Middle Eastern energy resources will require cooperation, which is based on mutual respect. An ideology that paints one of the world's great religions as a hotbed of violent radicalism will not endear us to the Muslims that form the overwhelming majority of Middle Eastern populations. Nor will invasions and occupation win us many friends. The Middle East is critically important in terms of its energy potential. The United States, if it is to have access to these resources, which under current conditions is unavoidable, must seek to be an acceptable partner for the region, with a positive image both among governments and among the general population as well. The current policies are producing the opposite result, generating wide-ranging hostility and complicating all future arrangements and negotiations.

Third, the considerable expense of the various conflicts is a major contributor to the U.S. budget deficit; this in turn has an adverse effect on the value of the currency. The currency depreciation in turn artificially raises the oil price in dollars, leading to inflation and other problems. This particular issue is also addressed in a subsequent chapter.

Finally, the "war on terror" is a huge distraction at a time when the country needs to concentrate on more pressing and real issues. Wars, even if artificially created and fundamentally unnecessary, have a tendency to hog national attention and to put "patriotic" ideology ahead of real national needs. Our focus needs to change, and for this the failures of "war on terror" needs to be recognized, and the attendant hostilities wound down as quickly as feasible.

Up to this point we have discussed the fundamental importance of fossil fuels in the birth and growth of the modern economy. Focusing then on oil, the most valuable of these fossil resources, we have outlined policies that would allow the U.S. to increase and stabilize its oil and fuel supply in the face of the gradually developing shortage and increase in prices. We will now turn to the domestic impact of this development and to the policies needed to deal with its consequences.

DOMESTIC POLICIES

The paradigm shift from growth to efficiency will result in a "new" economy. To achieve this in an orderly manner, we will first have to leave behind the old ideologies of capitalism and socialism. This will allow for the development of a new partnership between the markets and the state, based not on opposition but on using the respective strengths of each.

CHAPTER NINE

The New Economic Landscape

The reduced availability and rising price of energy, and specifically fossil energy, will have a major impact on economic life. The modern world has run on fossil energy for two hundred years, and still does. What we are dealing with in the short term is a growing limitation in the supply of the most flexible and useful form of fossil energy: petroleum. But such limitations do, and increasingly will, apply to other forms of fossil energy as well. There are still abundant reserves of natural gas and much larger ones of coal. These other fuels however cannot replace oil as far as its major uses are concerned, with transportation being first and foremost. In addition the prices of all sources of fossil energy tend to move in tandem, so that the rise in price of one will affect the price of the others, whether they are widely available or not.

In the end, these gradually developing limitations will totally transform our economy. As we have already mentioned before, if an essential commodity is in short supply there are three approaches to dealing with the shortage. In the case of oil, these approaches will run as follows:

- Produce more domestically, through increased drilling, new production methods or compatible substitutes such as synthetic fuel.

- Secure a larger and stable supply from abroad through development, for instance, of new resources in Russia.

- Extract more useful energy from what we have or can get, *by increasing the efficiency of use*, or making better use of the energy sources that are available.

At the beginning of this book we have explained how the original abundance of fossil energy sources gave rise to the paradigm of economic growth. Because such resources appeared to be limitless it was assumed that growth could continue indefinitely. We are now at the point where growth is bumping against limited energy supply. The era of pure economic growth is ending. The economic paradigm of the future will be *efficiency*.

Efficiency means getting more work out of the same energy source. Greater efficiency allows one to maintain the same level of activity or comfort with a reduced expenditure of energy, through the use of better technology. Increasing efficiency in all our activities will enable us to keep or even improve our living standards even as energy consumption drops.

The substitution of efficiency for growth as the primary economic paradigm will have the following benefits:

- It will allow us to maintain and possibly even increase economic activity over the next decade or two despite the gradual reduction in available oil supplies.

- It will provide the necessary time window for the development of new energy-producing technologies.

- It will shift our economic focus from consumption to investment, a necessary change of which more will be said later.

- It will relieve the pressure on the environment and reduce the carbon emissions held responsible for climate change.

This change will take place whether it is planned or not, as the underlying causes are beyond our control. But if we understand it and proceed with the necessary planning, we can minimize its disruptive impact and achieve a relatively smooth transition. This in turn will require the abandonment of the political paradigms associated with growth.

THE NEW POLITICAL REALITY

Since the Industrial Revolution the Western world has suffered from a split personality in the area of economic thinking. This problem originates in the divide between Right and Left, which, as stated earlier, is grounded in the political choices stemming from the Industrial Revolution.

The credo of the Left has been that the economy must be "fair" and that everyone should receive a more or less equal share of the national wealth. For this to be possible the state must be the designated owner and manager of the means of production. In other words, in socialist ideology a "fair" economy is one that is (a) nationalized, and (b) controlled and managed by government officials. This doctrine equates free enterprise with greed and private ownership with the exploitation of labor. By contrast, the Right claims that state ownership amounts to slavery as well as low productivity, and that the profit motive is the chief driver of efficiency and innovation.

The Left-Right struggle is, in actuality, basically over; capitalism and socialism have blended to form various hybrids which partake, in varying degree, of both ideologies. Despite that historical evolution the *ideologies themselves* are still very much alive and kicking in the minds of their supporters, regardless of the objective facts. The standard-bearers of the Right are still vigorously promoting the benefits of unrestrained capitalism, while their intellectual adversaries on the Left still insist that the state must be the provider of last resort. The ongoing debates over the pros and cons of globalization or of government-provided health care are ready examples of the fact that neither side is willing to sacrifice ideological commitment and purity for the sake of either objectivity or effectiveness in resolving issues.

The decay and collapse of the socialist economies has clearly demonstrated that the state is *not* a good provider and manager. We have also seen, particularly in the case of Russia, that the introduction of wild capitalism after socialism collapsed did not produce general wealth, but instead accentuated social and economic disparities between the mass of the people and the new economic elite. The condition of the average Russian under the new capitalist "oligarchs" was not substantially different from that of the workers exploited by the nineteenth century robber barons.

In truth neither capitalism nor socialism can, in their pure form, produce satisfactory results. It is, therefore, worthwhile to reach beyond standard economic ideologies and establish what either the free market or the state is actually *good* at.

State management of the economy is inefficient because planners are not held responsible for the consequences of their mistakes, and also because, being functionaries and/or politicians rather than technical specialists, they generally do not have the competence and skills economic management requires. The state does have, however, compensating advantages: it can take a long-term, strategic approach, and at the same time has the legal authority to impose its views. The main motivation of the state, not only in theory but quite often in practice as well, is the general good of the nation and the protection of its interests. The state or government cannot manage, but it can *plan* and *direct*. Overall direction and inspiration is what we expect of our leaders, be they hereditary monarchs or elected representatives. The more difficult and urgent our situation is, the more direction we expect from the top, and the more of our independence and goods we are willing to surrender so as to allow the government to assume leadership.

On the other hand, the free market excels at generating efficiency and innovation, simply because personal profit provides a direct reward for these. But the power of the profit motive is at its maximum in the short term, when the effects of one's actions are most predictable. It fades quickly as the time horizon is extended and for that reason it is extremely difficult for participants in a free enterprise system to think strategically, unless they are supported in doing so by state power and resources. The longest commonly accepted time window for calculating return on investment is five years, and in most cases it is much less. An executive who is willing to lose money for five years for the sake of a projected later benefit will not last long in the free market system as it exists in the U.S..

Yet that five-year span is a short one when national interests are taken into consideration. In practice, the only private corporations willing to think and invest beyond the five-year horizon are the oil majors, and this makes "Big Oil" into a major national asset in the current energy crunch.

The response to a major long-term national need or emergency requires strategic continuity, which is best provided by the state. It also demands the innovation and creativity that are naturally provided by free enterprise. The energy situation we are now entering can be termed to be a national emergency. Total state control or ownership will not provide a solution, but neither will the markets if left to themselves. What will be needed, and what will eventually occur because of sheer necessity, are multiple forms of public-private partnership. This is a new political paradigm, which contradicts, and will soon displace, the old ideologies of Right and Left, which soon will no longer be relevant or useful.

From this need for cooperation will develop a new and unfamiliar economic landscape. The following sections give a glimpse of what is to come.

CHAPTER TEN

The Fading of the Consumption Ethic

As already stated, the seemingly limitless supply of energy achieved through the use of fossil fuels has generated a parallel belief in the possibility of indefinite economic growth. This has led to a major difference between modern economic thinking and the mentality of preceding periods. Before the Industrial Revolution, wealth was seen as the accumulation of durable assets: precious metals, land, tools and weapons, long-lasting buildings, and so on. Over the last two hundred years this asset-based concept of wealth was displaced by one based on *turnover*, in which economic activity itself became the new standard, regardless of what was actually produced. Income replaced assets as the main criterion of economic success. Once the state learned how to tax income it sought, just as the capitalists themselves, to maximize this new source of revenue. At that point the *growth* of economic activity, by common consent, became the nation's goal.

Put in the simplest terms, the increase in economic activity is the product of the increase in market size multiplied by the increase in the purchasing power of each active entity in that market. In other words, to achieve maximum growth the market must expand to every individual in the targeted population, and simultaneously the purchasing power of each individual must be maximized. The first goal will be achieved if every individual in the market becomes a buyer, or in modern parlance a consumer. The second will be attained if everyone has at their disposal an unlimited amount of money. This is not fully feasible, but can be approximated through a financial system oriented toward generous lending. Thus, an economic policy focused on growth will inevitably result in a consumer economy driven by credit.

If the energy supply becomes limited, the above model becomes not only unpractical but undesirable as well. At that point there will be no choice but to either reduce the size of the market or the consumption of each individual. If one wishes to avoid a Darwinian struggle through which the strong overcome the weak in order to keep the lion's share of the available resources, the goal of increasing consumption must be abandoned as a driving principle and replaced by another paradigm.

When the economy is expanding in quantitative terms the main social and political concern is to ensure that the newly created wealth is, as far as possible, evenly spread through society, according to the principle that a rising tide lifts all boats. That some boats, particularly those of the top income group, rise faster than others is acceptable as long as wealth disparities are tolerated by the majority. When that ceases to be the case the government has to redress the balance. This is the political logic behind the introduction of the graduated income tax in the early twentieth century, which marked the end of a period when the U.S. economy had been expanding rapidly for two generations, producing a huge accumulation of wealth at the top of the income pyramid.

In a situation where continuous growth is no longer possible, the main goal will shift. Because the size of the economic pie no longer increases, the risk is that the least favored sections of society will drop through the subsistence floor into outright misery, a situation that generally leads to political disorder and to social upheaval. The main concern of government will then be to put a floor under the minimum incomes. Such a policy was in evidence during the Great Depression through the creation of a number of make-work schemes and of the Social Security safety net.

Up to now such policies were implemented primarily in the troughs of economic cycles. As the reduced availability of energy gradually crimps growth they will become a permanent feature of government activity. At the same time there will be a parallel shift in economic policy from encouraging maximum consumption to ensuring the satisfaction of minimum needs. In the field of transportation, for example, this means that rather than encouraging everyone to buy the biggest feature-laden SUV, the priority goal would be to ensure that everyone has some means to get to where they need to go. This is *not* meant to translate into some form of universal rationing, coupled with a prohibition to consume above a set level. Human society does not function that way, except in the direst emergencies, and even then not for very long. It does mean, however, that the first priority will be for the basic needs of every citizen to be covered to the extent that is practically feasible. Beyond that, some determination will have to be made concerning the disposal of the available economic surplus.

It will be said immediately, in response to the above, that such a scheme can only be made to function through a significant intervention of the state, which means collectivization and socialism. That it requires some state intervention is true; that it leads to something like communism is not. We already have similar schemes in the U.S., such as Social Security, food stamps, rent assistance, and other such. Their aim is very similar to what is outlined above, but they are not part of a plan to create a socialist state. What is likely to happen in the coming years, with the energy crisis providing additional urgency, is that these schemes will be rationalized and integrated into a more coherent and efficient whole, with some new services, such as basic health care, thrown in. But the objection concerning a possible return to socialism is a valid one and needs to be examined and answered at greater length, for it points to the other major change that the energy crisis will bring about: the changed relationship between the market and the state.

CHAPTER ELEVEN

The State and the Market

Of the two economic ideologies born of the Industrial Revolution the first, capitalism, demands maximum entrepreneurial freedom as well as complete reliance on market mechanisms, while socialism dictates a uniform distribution of goods within a state-controlled economy. While the two concepts have given rise to politically adversarial camps, realism has also led political leaders on both sides to recognize *that neither one proved sustainable in its pure form.* Those who refused to recognize that fundamental truth and adapt their views accordingly, such as the leaders of the now defunct Soviet Union, eventually fell out of power and were replaced by others more willing to compromise. By now the differences between the various blends of socialism and capitalism became small enough to be politically trivial for anyone except committed ideologues.

It is, therefore, high time that the "Right vs. Left" political paradigm be finally discarded and replaced by one which better combines the functions and strength of free enterprise and state power. Such an attitudinal change will allow for the restoration of a normal relationship between the markets and the state.

A market economy, driven by the pursuit of short-term profit, has always been and still is the most efficient agent for the production of goods and achievement of technical innovation. The state, on the other hand, can take the strategic view and use its power to ensure that the nation's resources are used in accordance with its long-term interests. The need for increased state intervention is proportional to the urgency with which these national interests must be defended or promoted.

For this reason the population usually accepts, and often expects, the extension of state power in time of war or national emergency, while finding it an undue burden in times of peace and plenty.

State intervention in the economy is not, by nature, either outright evil or counter-productive, provided it is properly calibrated for the circumstances. Nor should it necessarily involve the nationalization of enterprises and the establishment of permanent state control. Whatever the ideologues of the Right and the Left might claim, effective collaboration between government and private enterprise is entirely possible. A prime example of such a necessary, and also extraordinarily effective, collaboration between the state and private industry is provided by the phenomenally successful U.S. industrial mobilization during WWII.

Not only did this effort achieve, in terms of technical innovation, logistics planning and production, results hitherto considered impossible. It also resulted in a major strengthening of U.S. economic initiative and industrial power. For the duration of the war, the U.S. government exerted nearly total control on the allocation of national resources and the use of the nation's industrial base. Yet no industries or sectors were nationalized and wartime controls were removed as soon as their purpose had been achieved. The public-private cooperation during the war completely revitalized the U.S. economy, and created an industrial complex of unprecedented power and reach. It also laid the foundation for the subsequent expansion of American corporations over the entire globe.

There are definite similarities between the challenge posed by WWII and the one presented by today's energy situation. It is logical to expect that answering the present challenge will demand a similar cooperation and partnership between government and industry, with analogous setting of priorities, allocation of resources, and control of key economic functions. It is also correct to assert that government control *alone* will not suffice. Government control in isolation from market input and the methods of private industry will in fact lead to failure. It is imperative in such situations that plenty of room be left for the initiative and inventiveness that are characteristic of free enterprise; for the private profit as well as for the operation of supply and demand; and finally for the setting of prices by the market. But there need not be any contradiction between the respective roles of the state and of the private sector. The general direction is set by the government and the implementation is the domain of private industry. It has been done before, and it certainly can be done again.

In order for such cooperation to be effective, however, the appropriate context must exist. Today there is a major difference with respect to the WWII period. At the time the United States had full political and economic sovereignty. Today the political sovereignty is still there, but part of the economic sovereignty has been surrendered, both to international institutions and to the anonymous phenomenon referred to as "globalization." The U.S. is no longer master in its own economic house, and it is to this issue that we will turn next.

CHAPTER TWELVE

The Dead End of Globalization

The economic globalization of the last two decades is both the culmination and the swan song of the unlimited growth paradigm. The political "victory" of capitalism, as demonstrated by the fall of the Soviet Union, has led to a strong trend toward the removal of all controls on the international flow of capital and goods. The United States has been the major leader in this development.

The motivation was for international corporations to gain the opportunity to seek the lowest possible costs of production and the consequent increase in profit margins. This goal has been achieved, temporarily, but at the expense of the transfer outside of the U.S. of entire sectors of the economy, due to a major discrepancy between theory and reality. The theory has been that, once all barriers are removed, the global marketplace would become an open and level playing field where the market would create the maximum economic efficiency in the production and distribution of goods and services.

In reality the trend toward a global marketplace has not produced the promised "level global playing field." While some states have practiced genuine free trade, others have taken advantage of the new situation to practice unabashed economic nationalism: the manipulation of customs duties, capital controls, currency exchange rates, taxes and subsidies so as to accumulate wealth and concentrate economic activity within their own borders. The consequence of this development was to allow the states implementing such policies to achieve huge trade surpluses at the expense of "free trading" nations, which by contrast lost a large amount of industrial activity by exporting it abroad. The impact of such transfers in the field of energy consumption has been major.

Globalization and the resulting transfer of industrial production have increased global energy consumption in several ways:

- First is the impact of the increased global shipment of goods, from heavy bulk commodities to finished goods. This increases demand for transportation fuels, all of which are petroleum-based.

- Second is the transfer of production to countries with lower labor costs. These tend to operate at a lower level of efficiency in terms of energy use, so that the transfer of economic activity increases the energy content of the goods produced.

- Third is the fact that the beneficiaries of globalization are rapidly increasing their energy consumption, not only for export manufacturing and the related import of raw materials, but to support rising internal consumption as well.

The globalized economy might be, in the short run, more profitable financially, but it is more energy intensive than the national economies it is in the process of replacing. The rise in energy costs will inevitably overwhelm any savings achieved through the use of cheap labor, rendering the entire process not only futile but increasingly harmful as the cost of energy rises.

The combined effect of these factors has been to bring forward the moment when the paradigm of indefinite growth would collide with the finite nature of fossil energy resources. We have seen an early example of this in the energy price bubble of 2007–2008. In that case runaway demand from export-oriented developing economies collided with the global oil production ceiling, causing an artificial shortage and a rapid rise in price. The price increase, amplified by massive financial speculation, walloped the consumer and has been a significant contributor to the current worldwide economic slowdown. This globalization episode thus has, unfortunately, provided a perfect illustration of the limitations of the growth paradigm.

The above means, in practice, that globalization has reached its peak, in great part because it is physically running out of fuel. There is, however, another reason why the reversal of the globalization trend will continue. We have stated above that the criterion guiding economic policy is shifting away from the promotion of quantitative growth and toward ensuring the satisfaction of basic needs. This implies some degree of state oversight of the production of basic necessities, which oversight can be effective only when such production takes place within the national territory.

In this respect the recent spike rise in oil prices, caused by runaway growth rates, was the first warning bell. The rise in food prices was the second. Globalization has given us, for a few short years, a world awash in cheap goods. It has also led to huge fluctuations in the price of essential commodities. No government intent of maintaining basic living standards can afford such a situation.

The move to globalization was driven by cost, particularly the labor cost involved in manufacturing. Energy was left out of the equation, on the old assumption that it would remain plentiful and cheap. That assumption has proven to be obsolete. The new economy will increasingly be driven by energy efficiency and the need for local control over production and distribution. The shift demands a pull-back from globalization, and the sooner this occurs, the lesser will be the economic impact of the "energy crisis."

The impact of globalization however does not end with the "real" or tangible economy. The financial side has been hugely affected as well.

"REAL" ECONOMY VS. FINANCIAL ECONOMY

The economic growth paradigm requires constant increases in consumption, which are made possible through the provision of cheap and widely available credit. Since credit can easily be expanded beyond the borrower's ability to pay, the consumer economy will always be unstable, alternating between phases of expansion followed by contraction, between boom and bust. Such oscillations have been a normal feature of capitalist economies throughout the industrial age.

Another consequence of the importance of credit in a growth-based economy is the size and importance of the financial sector. As this sector grows larger it tends to acquire a life of its own, separate from that of the "real" economy; "real" being here understood as referring to the portion of economic activity that deals in tangible assets, products and services. Once the financial world has reached a certain size, it can begin to create its own internal wealth, based on assets or instruments the value of which is based more on *perception* than on any direct connection with tangible collateral. The uncontrolled creation of such purely financial assets generally leads to "bubbles" which tend to amplify economic oscillations and in the end cause panics and crashes.

This capacity of the financial system to "over-inflate" has been vastly increased by globalization and is to a great extent responsible for the financial crisis that started in mid-2007. Under the global free market theory the removal of all barriers to, and controls over, the movements of goods and capital allows economic activity to seek and find its optimum location where goods and services can be produced at the lowest cost. The most globalized part of the world economy is finance, where enormous sums of money move instantly around the planet, looking for optimal returns. In theory, this is ideal, and indeed it has allowed for huge profits to be made. But, as always, there have been unintended consequences.

Due to the fact that the dollar has been the world reserve currency, the globalization of financial markets has allowed the United States to run increasingly large trade and budget deficits, relying on worldwide borrowing to cover them. In theory these outflows would be recycled through foreign investment in the United States. But in order to absorb such investment the U.S. economy must maintain a rate of growth corresponding to the increase in the size of deficits. This is difficult when large segments of the economy are being "exported" to lower-cost countries. U.S. dollars are, therefore, accumulating abroad, creating a huge floating monetary mass.

This monetary mass has grown even as the pool of U.S. assets available for investment has been *shrinking*: entire industry sectors are being outsourced; American corporations have been buying back large quantities of their own shares; companies have been taken over by private equity operators and taken off the market; and several categories of investments are now off limits to foreign companies due to security concerns. Over time the U.S. deficits have created a growing imbalance between the demand for investment opportunities inside the U.S., and the actual supply of such opportunities. This imbalance has had two untoward consequences.

First, even as investment opportunities in the "real" U.S. economy were thus disappearing, enormous amounts of money were still flowing into the financial system to be recycled. As always happens, supply has risen to meet demand; since the real economy could not absorb them, the excess funds were diverted into the "shadow market," where demand for high-yielding investment generated a corresponding supply of new financial instruments and assets. These were both increasingly complex and opaque, and had only a tenuous connection to any kind of tangible collateral.

The worth of these new "assets" depended to a great degree on a *perception* of value rather than on worth that could be objectively measured. Sooner or later this assumed value was bound to be questioned. This began happening in mid-2007, and it has since been found that the value of such assets is either impossible to determine or simply nonexistent. For the institutions holding, or suspected of holding, such assets, the resulting situation is somewhere between uncertainty and insolvency. The unknowable worth of many balance sheets has led to the collapse of the confidence that financial institutions have in one another, the breakdown in lending, and the resulting credit crunch. We now have a financial system that is highly dysfunctional because it does not know its own worth and, therefore, is unable to extend credit.

Second, the "floating monetary mass" referred to above is so large that its movements will swamp any control measures set up by national governments. It moves across the globe without barriers or restrictions, seeking the highest returns, and creating "asset bubbles" wherever such returns are perceived to be available. The housing and commodities bubbles we have just experienced are excellent examples. The moment any asset class, such as oil, shows sign of increasing in value, the global money will descend on it and prices will once again, for a short time, rise to improbable heights. Such asset price spikes can totally dislocate an economy and destroy both business and public confidence.

There are two ways to reduce this problem to a more manageable size, and set the economy back on a sound footing. One is to stop feeding this "global monetary beast" by reducing our deficits, and to take measures to discourage large flows of speculative capital. These will be dealt in a later chapter. The other approach is to create new and real investment opportunities that will soak up and retain the excess liquidity. This is where the energy issue provides a way out.

As energy becomes more expensive due to limited supply, growth will be replaced by efficiency as the primary criterion of economic performance. Consumption then ceases to be desirable *per se*, and is reduced to what can be termed "useful consumption," which is directed first at the satisfaction of basic needs. The priority shifts to *investment* and primarily to such investment as will increase the general efficiency in energy use as well as the energy supply. This completely changes the financial picture. Within an economy that has previously been dedicated to maximum consumption the potential for energy-saving and energy-producing investment is almost infinite. Investment requires the *accumulation of capital*, rather than the expansion of credit, and thus favors *saving over spending*, which turns our current economic ethic on its head.

The current economic downturn, by its timing and its very severity offers the opportunity to initiate the change outlined above. So far all measures taken by the authorities have aimed at "jump-starting" the economy through massive deficit spending and increases in financial liquidity. While some of these measures might be useful to a limited degree, the overall approach is bound to fail in the long term. The deficits and greatly expanded money supply only add to the global money pile, guaranteeing the intensification of boom-bust cycles down the road. The associated price spikes will destroy confidence and rapidly quench any revival of consumer spending.

This general shift in emphasis will favor tangible investment over activity that is essentially financial. The needed control of resources, which will be proportional to the urgency of the situation, leaves little for purely financial activity. Activities of a speculative nature, such as bets on price movements or trading in commodity futures, are likely to be discouraged if not outlawed outright, as often is the case in wartime or other such exceptional situations.

One can conclude from the above that a situation of energy scarcity will shift the focus of economic activity back from the financial sector, where it is focused today, to what is commonly called the "real economy." The bulk of investment will be channeled into the creation of tangible assets, such as infrastructure, production facilities, scientific research, and engineering, rather than into the production, trading, and manipulation of financial instruments. Because of the need for economic control by national governments, international financial activities will be strongly curtailed as well.

Enough has been said to suggest that the future "new" economy will differ in significant ways from the one we have known. The crisis that began in 2007 might yet demonstrate that the economy we have known, based on the growth paradigm and on globalized financial flows, is no longer functional. What would take its place needs to be sketched out, which will be done in the following chapters.

CHAPTER THIRTEEN

Redefining Roles

In the previous chapter we have outlined the functions of the free market and the state: the state directs and the market produces. If we are to move as quickly and smoothly as possible from an economy driven by consumption to one driven by investment aimed at energy efficiency, we need to define a context within which these respective strengths will optimally combine and interact.

Regardless of how it is empowered, democratically or otherwise, the government works top-down. This mode fits its strategic vision and its leadership role. By contrast, free enterprise works from the ground up. The greatest potential for initiative and creativity resides in the individual, who can choose to save or spend, to invest or to consume. The larger and more hierarchical the organization within which the individual works, the less space and opportunity there is for initiative and creativity. The greater the size of a business organization, the more likely it becomes that, at the leadership level, the economic mission will become subordinated to the ambition to reach the top spots, with their huge financial rewards and extensive privileges. Large business organizations are neither nimble nor innovative. One needs only to look at the current economic debacle to become convinced of that.

This does not mean that all businesses should be small. Large corporations have their place, but they are best suited for steady activity that does not require much creativity or innovation. They are, therefore, best suited to carry out missions of a strategic nature and stretching over long periods. In the situation we are concerned with here, the shift from a consumption economy to an investment-based one, where extensive government direction will be needed, large corporations are most likely to fall under close government control. "Control" here does not mean *ownership*. It means the corporation basically does what the government tells it to do, like General Motors did during WWII.

By contrast a small business organization needs to be both flexible and creative. It must adapt in order to survive and innovate in order to grow. Every talent within the organization must be nurtured and exploited, and participation by all members is needed. It is no wonder that in the U.S. today small businesses generate 75 percent of the new jobs and over 90 percent of the new ideas. It is at ground level that innovation and creativity are most intense, and they become less so as the business grows and acquires the familiar pyramidal shape.

In a situation such as we are now entering, where a major shift in economic paradigms is taking place, it is essential to create a context within which strategic direction and creativity combine as close to the ideal as possible. This will mean relatively strong state control at the top of the economy, and maximum flexibility at the bottom, with the hand of the state getting lighter as the organizations get smaller. At ground level there should be near-complete freedom.

In an earlier chapter we have brought up the case of the WWII industrial mobilization as an example of how the state and private industry can successfully cooperate, without falling into the trap of nationalizations and thereby drifting into socialism with all its inefficiencies. The key to making such cooperation a success is, first, an understanding of the respective strengths of each member of the partnership. A second essential element is a clear concept of why the cooperation is necessary and how the goal is to be achieved. In WWII this was clear from the moment the Japanese navy attacked Pearl Harbor. The goal was victory over the Axis powers, and the means to that end was the mobilization of all U.S. resources for the production and delivery of the necessary armaments, fuel and other supplies.

The goal of achieving a stable and secure energy supply, together with top efficiency in energy use, will require a less intense effort than a major war, but it is equally important for the long-term survival of the nation, its culture and its institutions. This effort will also stretch over a longer period than WWII and will have to be maintained over several presidential administrations. Because of this longer time span, it is critical that a national consensus be built from the start, and that the requirements of the task are well understood. Only then can a "new economy" be built without major disruptions occurring in the country's social life and body politic.

The first prerequisite for building the new economy is, as stated above, the abandonment of the ideological enmity between the state and the private sector. This will require, on the part of the state, a renunciation of its role as *provider*, derived from the socialist mode of thinking, and an emphasis on its role as *leader*. Here it must be understood that leadership is not a position but a function. A person does not become a leader by being born or elected to high office. Effective leadership requires, first, a correct understanding of the national interest; second, the development of a strategy to successfully pursue such interests; and finally, an ability to unite and organize the people for this very purpose. If the goal and the strategy are not correctly defined and unity of purpose is not achieved, all effort and resources dedicated to the task will be wasted and the nation will end up being worse off.

The private sector, on the other hand, and this particularly at the highest levels, needs to abandon the capitalist dream of *total freedom*, and focus more on the ability of free enterprise to *efficiently execute designated tasks*. Whereas the state, with its strategic and long-term vision, can define *what* needs to be done, it usually fails in the *how*. How to get things done efficiently, on time and at the lowest cost is where the private sector excels. From this it follows that the most effective government, particularly in emergency situations, will be a *small* government, tightly focused on the goal to be achieved and the attendant strategy. Only those economic components that are essential to the execution of the strategy need be under government control. Everything else must be left to the private sector, so as to ensure that creativity, innovation and efficiency are given free play.

This is the same assignment of roles that led to the extraordinary success of WWII mobilization. It must be noted that the government, which ran the greatest war in history was, by current standards, extremely small. Nevertheless its authority went unchallenged, its reach extended over much of the globe, and its ability to marshal national resources was unequaled, before or since.

The task today is not victory in war, but it shares some of its characteristics, since it also pertains to our national, economic and cultural survival. Losing the "energy war" would leave the United States in a greatly diminished state, and force it to surrender much of the amenities and freedoms we now take for granted. In order to ensure the final success an effort similar to WWII mobilization will be needed. How shall we structure our economy in order to maximize our chances of success?

The "new" economy will be defined by the co-existence of, and cooperation between, private and governmental economic activities. In the current situation the state and private enterprise interact, and often conflict, at every level, giving rise to considerable inefficiency and confusion, as well as to pernicious elements as lobbying, subsidies, speculation, and increasingly complex tax codes. The new economy will be structured that both the state and the private sector can have the maximum positive impact in their respective spheres.

- The key area over the next several administrations will be the energy sector, and that is where the government will have the greatest influence. This does not mean the U.S. government should nationalize and manage the energy industry as well as public utilities. In fact the state should own and manage as little as possible. Its task will be to outline, together with representatives of the private sector, a realistic strategy to ensure the nation's energy supply and to realize the highest possible efficiency in its use. Once this strategy is laid down, it must assign to the private sector the various tasks involved in the strategy's implementation, including research, development of new sources of energy, the creation of production facilities and the management of the energy distribution network. This will inevitably involve a fair degree of control about what needs to be done, by whom, and at what price.

- Connected to this core of energy production and distribution will be a number of economic sectors that are large users of energy and highly dependent on it, the most important of which is the transportation sector. A degree of government supervision over these sectors will be needed in order to set efficiency targets and ensure their attainment, but such control will be less direct than in the core sector and allow for more leeway as to the way the targets are reached.

- Further out are the numerous industries and businesses that still use significant amounts of energy, but for whom energy is no longer the primary input. Here again a constant improvement in efficiency will be needed, but these will be through market mechanisms and incentives rather than through direct control and directives, so as to leave increasing room for innovation and private initiative.

- Finally, at the periphery, are individuals and small businesses. Here market mechanisms and the law of supply and demand will be most effective, and state intervention will be least effective. Government directives and controls should, therefore, be reduced to the minimum needed, or eliminated altogether.

What are the critical differences between what is sketched out above as "the new economy" and what we now have?

First, in the above sketch of the "new economy" the respective roles of the market and the state are focused in the areas where their respective advantages are greatest: government influence and control rise toward the top, where a strategic approach and a long-term vision are needed to serve the national interest. But the "heavy hand of the state" is withdrawn from the areas where private initiative and innovation work best.

The differences between the current and future economies outlined above reflect the true importance of energy resources, and in particular the impact of the availability of petroleum, on the functioning and management of our economy. The phrase "energy crisis" is often used to describe our current situation, but this designation is incorrect. A crisis is, by definition, a short-term and transitory occurrence, a disturbance in an established and enduring process or pattern. Such a development is usually precipitated by unexpected events; it is then adjusted to, dealt with, and resolved. The "oil shocks" of the 1970s could be in aggregate called a crisis. They were originally precipitated by a war no one expected to occur: the Yom Kippur conflict between Israel and the surrounding states. They surprised us and required both political and economic adjustments, but they did not modify the basic energy supply and demand situation. Petroleum supply was more than adequate to meet demand, and our fossil-fueled economy quickly recovered and continued to expand. Today, by contrast, we are dealing with a fundamental supply issue with major long-lasting consequences.

Second, the differences between the "future economy" sketched out above and what we know today illustrate how far we are, politically speaking, from the measures and policies required to manage our economy during this time of transition. For the current crop of politicians, the issue of oil and gasoline prices does not even rise to the crisis level. It is merely an election year problem, to be dealt with in the usual ways: subsidies to win the necessary votes; committee hearings to pinpoint convenient villains, with "evil" Big Oil in the forefront; and pronouncements on energy security and reduced dependence on imported oil, which are seldom if ever followed by effective action. None of that, of course, will change the situation one bit, except that it will get slowly, or rapidly if unforeseen events occur, worse as time is wasted and nothing is done.

The question then arises: What *is* to be done? Having sketched out the big picture, we will look in the next chapter at some of the initial and necessary steps.

CHAPTER FOURTEEN

Getting Started

In the foreseeable future, energy policy will be fundamental to national prosperity and even survival, and it is difficult to forecast at this time where it will end up. But it is possible to suggest a set of policies that can be implemented immediately and will not only relieve pressure in the short term but also point us in the right direction. Here are some meaningful first steps:

A. Develop a National Consensus

We will not solve this problem by "leaving it to the market" or "letting the government" take care of it. The required changes are so far reaching, the needed attitude shifts so radical, and the cost so great that anything short of a determined national effort will, at best, achieve only partial success. Therefore, the first item of business for any current or future administration is to recognize the magnitude of the issue; second, to define it in political and economic terms, as we have done in previous chapters; and following that, to start building the national consensus needed to deal with it. This means recognizing the need, estimating the costs and impacts of different policy alternatives, and defining a solution, or at least the beginning of one, that is both technically and economically sound. The proposed policies must then be presented to, and debated with, the electorate, so as to obtain their buy-in and future support.

Such a consensus does not exist now, and it will not be built in a day or in a year. An adequate set of policies will probably require several years to be fully developed and adopted, and that by itself will be a major achievement. The country will then be aware of where we stand, what the problems are, and what can be done to solve them. We will then have a common goal and be motivated to reach for it. From that point on the administration and the Congress will be able to proceed on the basis of the new consensus.

While the task of developing a national approach to energy supply is a daunting one, the alternative is worse. The current laissez-faire approach, or drift, has only made the problem worse. Fuel prices are oscillating between the cheap and the unaffordable; supply is gradually getting less secure; government-induced demand for ethanol is contributing to price distortions and food costs. Last but not least, our rising trade deficit, caused in great part by oil imports, is hastening the eventual devaluation of the dollar. None of the current policies has any reasonable chance of making a dent in the problem, while the probability that the current course will worsen our predicament is quite high: the current course has in fact landed us where we are now. There seems to be no alternative to facing up to reality and to dealing with the issue, and for doing this effectively, a national consensus is indispensable.

B. Remove market distortions

The market by itself will not solve the energy problems we face, but market realities must nevertheless be recognized. Every way of providing physical energy, be it traditional, alternative or basically new, has a cost, and decisions must take this cost into account., therefore, the second step toward a sound national policy will be to remove existing price distortions, and this means the elimination, as rapidly as possible, of all energy subsidies and legislative mandates that currently affect the U.S. market.

In theory at least, subsidies for specific sources of energy are provided by the government in order to overcome the cost disadvantage of a desirable source of energy over a less desirable one: for example, to favor the development of wind power vs. power derived from coal. The idea is that if subsidies allow for the mass introduction of the desired power source, improvement in the technology and economies of scale will eventually "level the playing field" in its favor.

In fact the opposite happens. Providing the targeted energy source with an artificial cost advantage and/or a legislated market *discourages* the innovation that could make it truly competitive. Why would the owner of the energy source spend the subsidy on development when he can pocket it immediately as profit? Once assured of a market he will prefer to use existing and "safe" technology rather than take risks with new developments. The net effect is to "freeze" the subsidized technology in its current state and to direct the primary effort of the sector toward the maintenance of the subsidy rather than toward technical and cost improvements.

Subsidies thus create political lobbies that in the end have little to do with either cost or energy efficiency. But there is a more fundamental reason for avoiding them in the area of energy. Market cost is a good equivalent for the amount of energy it takes to create, install, operate and maintain a particular energy source, such as a power plant, over its useful life. To artificially lower these costs through tax-funded subsidies will result in the replacement of efficient energy sources through less efficient ones.

The proper place of government assistance is not in supporting the operation of energy sources, but in *research and development*, particularly in basic research and in the competitive evaluation of technologies. In the case of electric vehicles, for instance, government funds would be far more effective in supporting research into more efficient batteries than in subsidizing the purchase of electric cars which are currently not competitive in price or performance.

If a specific component of our energy supply cannot be provided economically by industry, then that component should be returned to the research lab for further development and testing. The government should contribute to the expense of such development in proportion to the ultimate strategic value of this particular source of energy.

The elimination of market distortions will provide the basis for developing a strategic plan with different time horizons. The plan will first identify the energy sources and technologies that are currently available and can be immediately used to provide an adequate and secure energy supply. Beyond that, the plan will address how to improve the efficiency of these technologies. A third step would be to review technologies that are not yet in general use but offer promise if further refined and developed. The longest time horizon will belong to fundamental research aiming at the development of an energy supply that is sustainable in the long term and will support a gradual transition away from fossil fuels. We are here concerned only with the initial stage.

C. Exploit Existing Opportunities

Going back to the present, the next logical step is to recognize what already existing alternatives can be immediately identified and exploited. This will deal primarily with extending the supply of the forms of energy we currently use, so as to gain a breathing space for implementing longer-term planning.

We have already referred to a couple of such opportunities in a preceding chapter. One was to enter into a strategic partnership with Russia to develop its untapped oil resources, offering advanced exploration and production technology in exchange for a supply commitment. A second one was the development of a domestic synthetic fuel industry based on our abundant and highly accessible coal reserves. Both are useful alternatives that require only a political decision to be implemented.

Such activities can be initiated immediately as they do not require the development of new technology and can be carried out in partnership between the government and existing private entities and corporations. Private companies would provide the technical expertise and management skills, while the government would provide political backing as well as such funding as is needed to cover the strategic business risk. Benefits would flow to both parties. On the one hand our energy supply will be increased in the short term, which is a critical requirement in the current situation. We will also witness a nearly immediate improvement of our strategic position. Both developments would be of considerable benefit to the nation. On the other hand the private corporations involved will obtain additional business as well as the opportunity to widen their experience and expertise.

Here it is appropriate to add a word about the U.S. energy industry and the oft criticized "Big Oil." While the energy companies, be they global giants or nimble entrepreneurs, are not free from common faults, they also possess an extraordinary amount of expertise, both technical and managerial. They also have acquired the capacity, relatively rare in business, to think strategically, plan for the long term, and take huge, but calculated, risks. For over a century they have managed to supply, at competitive prices, petroleum-based products critical to our economy: fuel for all types of engines, lubricants, asphalt, and a huge variety of petroleum-based raw materials used in the production of anything from drugs to plastics.

Neither expertise nor past services rendered, in peace or war, should be forgotten or lightly cast aside. The oil industry remains a quintessentially American asset and its capabilities will be needed in the coming years even more than in the past. In a way the people of energy industry have made our lives so easy that we hardly notice their contribution. Times are coming when we will be grateful that they are around.

D. The Road Toward Efficiency

All the measures recommended above can be implemented immediately. They require no new technology, only political will, funding, and strategic insight. Their benefits will be two-fold. First they will provide a substantial boost not only to the economy but also to the body politic. Even the formulation of a clear and realistic policy will have an important psychological impact both here and abroad. Its implementation will contribute to stabilizing the price of oil and of other sources of energy. It will tend to level the playing field between oil producers and consumers and keep inordinate price increases in check. There will also be a positive impact in terms of domestic investment, job creation and business confidence.

At the same time, and no matter how quickly and efficiently these policies and related projects are implemented and carried out, their impact will still be temporary. Increasing U.S. oil and fuel supply, be it natural or synthetic, by a few million barrels a day over the next four to six years will greatly relieve the pressure on both supplies and costs. But it will not reverse the trend toward ever costlier, and eventually less abundant, petroleum. It will have to some extent the opposite effect, hastening the discovery and exploitation of the remaining undiscovered deposits, be they in North America, Russia, or anywhere else. We will be buying time now at the expense of losing it later. It is a necessary choice, but its limitations must be kept firmly in mind.

There is, as we have written earlier, no ready substitute for petroleum and its derivatives. Nor do we believe it possible to satisfy all our oil needs through the manufacture of synthetic fuel, which can only be an interim solution. Even if it became a permanent one, the fact that all fossil fuel reserves are finite will require that at some point we develop alternatives, both to our dependence on oil and to our current concept of indefinite economic growth. It will be a long transition, so the sooner we begin, the better.

We cannot and will not return to a fully sustainable economy in the foreseeable future. The technologies needed for this have to be developed and demonstrated, and then huge investments in them will need to be made over decades. In the interim the only practical solution is to pursue greater energy efficiency. This will require a gradual transformation of our economy and the associated way of life, which will flow out of the national consensus referred to above. It will involve major projects and programs involving both private industry and government.

THE IMMEDIATE NEED—TRANSPORTATION

The future shape and structure of our energy supply will be determined by both technical and economic factors. The economic determination will be based on real market prices of the various alternatives, after current market distortions have been removed. The technical determination will be based on feasibility and efficiency. It is too early to guess what the sources of energy will be. Research on post-fossil sources is at a very early stage, and the nation's creative resources have not been focused on the issue. The best we can do is to set the stage, and relieve the existing pressure so that future efforts can proceed in an orderly manner.

The highest pressure building up now is in the area of transportation, which is entirely dependent on petroleum products. There are two ways to reduce the energy consumed through transportation.

The first is to reduce in a general way the length of travel, particularly for goods. The fuel-related costs of transportation are directly proportional to the distances traveled, so the greatest savings in this area will be achieved through stepping back from the physical globalization of trade and industry, which relies on transporting ever more goods, materials, and people over world-scale distances. After transportation involving global distances has been reduced to what is truly necessary, transport on continental and smaller scales can be tackled. In each case one will need to balance the cost of producing goods locally against that of shipping them over a great distance. Fortunately most reductions will occur on their own under the pressure of market forces, as the cost of fuel rises and becomes an ever greater fraction of the delivered price.

The second approach is to make sure that the transportation we do need is as efficient as possible. This involves first the choice of the type of transportation used, such as the choice between air, rail, road or water. Second, it requires us to maximize the efficiency of the selected vectors or vehicles, including not only the engines but all other components and systems as well. In this respect the greatest potential gain is likely to be in automotive transport.

There have been largely ineffectual efforts to achieve automotive fuel efficiency through "mileage standards" legislation. This is the least effective method. Mileage standards impose an undue burden on automotive companies, forcing them to choose between producing small fuel- efficient cars and larger, less efficient vehicles. Small cars are economical but not very profitable to build, while large vehicles are fuel-hungry but offer much better profit margins. As a result, attempts to legislate the production mix creates an unavoidable conflict: the automotive company will either obey the letter and spirit of the law, reducing its profits and thus short-changing shareholders, or satisfy shareholders at the cost of finding ways around the law, as has been the case with the production of trucks and SUVs. The legislative approach is thus in conflict with the profit motive, a situation that inevitably leads to evasion, lobbying, lawsuits, and other forms of resistance. These in turn will eventually produce the effect opposite to the law's intention.

The other drawback of the legislative approach is that it is a restriction of individual economic freedom. If such standards are really enforced, which is unlikely in the U.S., users will be forced to buy a vehicle they do not want. Here the economic motive works much better. Once the fuel price rises above a tolerable level, most users will trade down in terms of size and up in terms of efficiency. They do not need a law to do that, as has been shown whenever the price of fuel has spiked above a tolerable level.

Once legislated fuel economy standards are discarded the remaining means to reduce consumption are rationing, taxation, or improved vehicle efficiency. Rationing is an administrative nightmare and should be considered only in extreme situations, as was the case in times of war or embargos. Taxation, or making the vehicle registration cost proportional to engine fuel consumption, has been used effectively in Europe for a generation, and has resulted in engine efficiencies much higher than those prevailing in the U.S. Introducing such a proportional taxation system in the U.S. today would, however, have a very uneven impact, penalizing both those who use their vehicles for business and drivers with lower incomes. In addition, raising taxes, of whatever kind, in the middle of an economic downturn, such as the one that began in 2007, is not recommended.

This leaves efficiency, where huge improvements are possible. A large number of known, tested, and immediately applicable technologies already exist in this field. Implemented together, these will result in large improvements in the fuel efficiency of automotive vehicles *without loss of performance or safety*. Such technologies include: conversion from gasoline to diesel, low-friction tires, aerodynamic design, electrification of auxiliary engine functions, more efficient combustion, electronic gear shifting, and so on. The Department of Energy, among others, has a list of these, together with data compiled from the numerous development programs and tests it has funded.

There is in fact an excellent justification for a large government-industry program aimed at the general, immediate and coordinated introduction of these fuel-efficient technologies across the entire U.S. automotive industry. If properly planned, such a program would begin yielding results within the three-to-five year vehicle development cycle, or sooner as that cycle itself can be shortened. Such a program would not require any "nationalization" or even increased government control over the industry, but only a genuine joint effort backed by some legislative authority, as well as funding for development and engineering. The U.S. had scores of such programs during WWII and they were hugely successful, often surpassing the ambitious goals set for them in terms of both delivery schedules and production quantities.

Such a program would achieve two ends. First, it would reduce U.S. fuel consumption by at least a couple of million barrels per day, while in parallel relieving the burden of high fuel prices for all consumers, who would now have access to vehicles of the highest fuel efficiency practically attainable. Second, it will provide a training ground for much larger programs of the same nature that will be needed in the future for the development of an energy-efficient infrastructure and economy. Its psychological impact, both domestic and foreign, will be immense. It will have considerable benefits in the areas of management, technology, and manufacturing; it would revive the U.S. automotive industry, which is currently spiraling downward; and it would generate substantial new employment. The cooperative government-industry part of the program can be fully implemented within four years, the term of one presidential administration.

Such a program would be focused on *currently available* technologies and fuels. While the benefits will be undeniable, it is likely to raise objections on the ground that such fuels are carbon-based, and that their use raises the danger of "global warming." The next chapter will address that subject, which is fundamental to any debate about energy.

CHAPTER FIFTEEN

The Issue of Global Warming

This is a book about energy, not climate. But as our current sources of energy are primarily of the fossil variety, the subject needs to be addressed in any comprehensive discussions of present and future energy issues.

In its simplest form, the "global warming" argument runs as follows:

Fossil fuels are primarily compounds of carbon and hydrogen, and originated when the remains of living organisms were trapped in geologic sediments and underwent a slow transformation into oil, coal or natural gas. The trapped carbon came from the earth's atmosphere, and the rapid burning of fossil fuel releases it back under the form of carbon dioxide, or CO_2. Carbon dioxide is a "greenhouse gas" which traps heat within the atmosphere, thereby warming it.

There is currently a broad scientific consensus that such warming is de-stabilizing the earth's climate, with mostly untoward effects: the melting of glaciers and of polar ice, with a consequent rise in sea level; an increase in extreme weather events such as droughts, storms and floods; the desertification of wide areas and the reduction in water supplies; a creep of tropical diseases toward higher latitudes; and a general weather instability, harmful to agriculture and other human activities. In view of the above there is a growing demand to reduce the use of fossil fuels as quickly as possible.

With respect to this issue we will briefly review the following:

— The status of scientific knowledge

— The current political impact of the global warming issue

— How the policies advocated in this book relate to the subject of climate change

A. Status of Scientific Knowledge

Scientific advances normally proceed through four stages:

- First is the *question*: facts are observed or discovered that cannot be explained through established scientific laws or existing theories. A new explanation must, therefore, be developed.

- The second stage is the development of a variety of such explanations or *theories*, each of which will tend to be backed or preferred by a different group of scientists.

- As research and discussion proceed, one theory gradually emerges as the best fit for the observed phenomena. This theory then becomes the *working hypothesis* around which most of the further research work will be carried out.

- Finally, a complete logical explanation, detailing the causality chain from initial cause to the observed effects, is developed. The definite test of the theory is its ability to *predict*, correctly, that if certain causes operate, then the specified effects will follow.

- The greenhouse gas-based theory of climate change is currently at the stage of the working hypothesis. It is accepted and supported by a majority of climate scientists and fits the observed facts reasonably well. Also in its favor is the fact that so far there is no viable competing explanation for the observed climate changes. The theory, however, has not been developed to the point where it would allow formulating *accurate predictions concerning future changes*. Climate is a complex phenomenon within which causality is still poorly understood. Predictions can, therefore, only be made in a very general manner and even then with reservations for the effects of still unknown or partially understood influences and processes.

- This uncertainty by no means invalidates the theory, but it creates a problem when it comes to practical planning and the elaboration of concrete measures for climate change mitigation. There is a considerable body of evidence showing that climate is indeed changing, but at this point there is a lack of detailed and reliable forecasts to which timelines and numerical goals can be anchored.

B. The Political Impact

Most of the attention and publicity given to the climate change issue is the result of efforts by two minorities: scientists and environmentalists. While the general public has been made aware of the issue through the media, it still ranks very low on its list of priorities. The growth of material consumption, requiring the use of fossil fuels, is still the guiding economic principle. To limit current consumption for the sake of a sketchily defined danger far in the future is not, at this point, a politically acceptable alternative. There are subsidies as well as promotion for "green" initiatives, but as explained in the previous chapter, neither subsidies nor legislation are likely to reach the core of the issue, which is our dedication to growth.

Where the general public stands, so do the politicians, who may pay lip service to "green" initiatives when speaking to specific constituencies, but otherwise ignore the issue or relegate it to the back burner. The same is true in industrial and financial circles, unless there is a "green" niche market to exploit, or a government research grant to apply for.

This state of drift is compounded by the fact that the potential effects of climate change, as predictions now stand, are difficult to distinguish from the normal vagaries of the weather. In addition, a number of destructive human activities, such as deforestation and industrial pollution, are far more visible and may obscure the effects of greenhouse gas accumulation in the atmosphere. The observed shrinking of the Aral Sea in central Asia, for instance, or the gradual drying up of the Yellow River in China are due mostly to misguided industrial and agricultural policy rather than to climate change, even if the effects of both could be similar in the long term.

This lack of perception of the climate change problem among the general public will only change if its effects become clearly visible and can be shown to be adverse to the interests of the majority of the population. If, for instance, icecap melting were to visibly accelerate, with cubic miles of ice shown sliding into the ocean on television, the issue of rising sea levels and its impact on residential patterns could be rapidly driven home. Barring such spectacular events, however, lip service and general indifference are likely to continue.

We are, therefore, faced with a conundrum: The existing scientific consensus as well as the evidence accumulated so far would make it prudent to take "human-induced climate change" seriously. But public opinion and government policy is still fixated on maintaining economic growth, which requires the continuing use of fossil fuels.

C. Connection to the New Energy Paradigm

To summarize what has been stated above:

- While the probability of human-induced climate change is high enough for states to pay considerable attention, effective mitigating policies are difficult to define and implement due to the current lack of accurate predictions as to the timing and specific impacts of global warming.

- Until such timelines and impacts are accurately known and can be verified, awareness of the issue within the *general* public will remain low and there will be little pressure for decisive action.

On the other hand, the energy issue is more visible and the public awareness is much higher: gasoline and home heating prices are a much more convincing argument than the prediction of potential disasters, however dire, that may or may not occur a decade or two from now. If policies aimed at resolving the energy issue could be coupled with those aimed at climate change, there would be considerable synergy.

If one assumes that economic growth will continue unabated *and* that the current climate change theory is correct, a reconciliation of the two positions is nearly impossible unless extreme pressure, such as the tangible and verifiable threat of major and near-term natural disasters, is brought to bear. If one assumes, on the other hand, that the reigns of fossil fuel and unchecked growth are, however gradually, coming to an end, the difficulty is somewhat reduced. The two forces of economic policy and greenhouse gas abatement would then be working in tandem rather than in full opposition to each other.

An economy that takes into account an inevitable reduction in the use of at least some fossil fuel will trend toward a lower carbon footprint faster than one where these fuels are still used to the fullest possible extent. An economic policy focused on energy efficiency, such as recommended in this work, would, while alleviating the problems caused by the reduced availability of fossil fuels, also mitigate carbon emissions and their potential effects on climate. Should the global warming hypothesis find further support and confirmation, we would then be already on the way to answering the challenge.

4

FINANCE AND BUDGETS

Even under the best of circumstances, the U.S. will be importing oil for another couple of decades. This requires that the value of the dollar be maintained, and that the main causes of its depreciation, our trade and budget deficits, be significantly reduced. With the proper mix of policies this is entirely feasible.

CHAPTER SIXTEEN

Fixing the Currency

Why is the currency important with regard to energy?

Even if all the policies recommended in the previous chapters are implemented on a priority basis, and a major effort is made to develop energy-efficient technologies, the U.S. will still be purchasing oil abroad for at least a decade or two. This is not a minor item. Our current import rate of roughly 12 million barrels/day translates into a yearly import bill of $350 billion at an average oil price of $80/barrel. This level is consistent with both the breakeven level of future oil extraction and with the budgetary requirements of petroleum exporters, so that it should provide the basis for future planning. Paying this year after year requires not only money, but sound money, and a financial system to match.

Before going further, a few words must be said about the "price" of oil, particularly in light of the wild swings of this price in 2007 and 2008. From a base price of roughly $45/barrel in January 2007, the price went to $145 in July 2008 before dropping to $40 in December of that year. This requires some explanation.

First this is *not* the price of all the oil traded on the planet. It is the "spot" price of a particular type of oil called West Texas Intermediate on any given day, quoted on the New York Commodity Exchange. "Spot" means that the oil is available for delivery. This spot market represents only a limited fraction of the total market for oil. The bulk of oil bought and sold is covered by long-term contracts, with fixed pricing or pricing formulas taking into the supply and demand situation so as to smooth out sudden variations. The spot market on the contrary tends to amplify supply and demand effects because of its much smaller size.

The price had been rising steadily between 2000 and 2006 because of increasing demand. In 2007 it spiked due to the following:

- Sharply increased consumption in China.

- Dollar devaluation: the oil price is in dollars. If the dollar lost value with respect to other currencies, which was the case, the oil price rose in direct proportion.

- Limitations in supply due to delays in major projects and production drops in a number of areas.

- Rampant financial speculation: 2007 and 2008 saw huge sums invested into oil by financial players who were not users of oil but wanted to ride the price rise in order to book paper profits.

The above factors combined drove the price to its July 2008 spike. Exactly the opposite happened afterwards: the global economy went into reverse; speculators unloaded their holdings; the dollar strengthened; and more oil came on the market. The spot price crashed just as it had boomed before. The drop from peak to trough was 75 percent, but actual oil consumption dropped only by 5 percent or so over the same period.

What is important is not the spot price but the underlying supply and demand situation which, as stated at the beginning of this book, is trending toward a gradual reduction in supply, and, therefore, a price that is rising over time. Large established oil fields, from which the bulk of world production comes, are declining at a rate between 6 percent and 9 percent per year. This requires new fields to be developed, but the bulk of "new oil" is not profitable below a range of $60 to $80 per barrel. If oil remains below that range supply will shrink relatively fast, essentially putting a floor under the price. Economic cycles will affect the pricing to a degree, but in the long term the price will rise.

In view of this trend and if all things continue as they are, the United States will be stuck with an ever growing oil bill. In many ways the U.S. currency problem is a circular one: the oil price is denominated in dollars, but as the U.S. sends large amounts of dollars abroad to pay for oil and other goods the dollar depreciates, which further raises the dollar price of oil and other imports; this causes the U.S. to export still more dollars, and so on. At some point, preferably before the country loses its current top credit rating, the spiral has to be broken and the currency put back on a sound footing. This is the fundamental issue that underlies everything else. The depreciation of the dollar has many negative consequences, but in the area of energy it can be fatal: worthless money means no oil, and a currency losing its value translates into an oil price that keeps on rising, dragging all other energy prices with it.

Before we go further, however, we need to open a parenthesis: the issue of the financial system, through which these huge sums are processed and moved around the globe.

THE FINANCIAL SYSTEM: TO FIX OR NOT TO FIX

In June of 2007 the global financial system began to collapse, causing a synchronized economic downturn around the world. As the downturn developed into a major recession a series of measures to "fix" the financial system were taken, on the assumption that if such a fix could be achieved, the affected economies would rapidly return to growth.

On the basis of this assumption huge sums, running in the trillions of dollars, have been spent, lent or pledged to a continuing effort of financial system repair, with very little effect as of the time of this writing. This raises two fundamental questions. First, can the system, as it now stands, be fixed? Second, should it be?

The background needed to answer the first question was provided in chapter ten. As explained in that chapter, there are two major problems within the globalized financial system as it now stands. One is the mismatch between the amount of money floating around the system and the investment opportunities available. The potential for investment in tangible assets is increasingly limited, but the supply of money continues to grow. The resulting demand for assets to invest in has generated a supply of purely financial instruments, the value of which is questionable and might never be determined, condemning the financial system to a perpetual state of uncertainty. At the same time, whenever a tangible investment area opens up, it is rapidly flooded with speculative money, creating financial bubbles that follow one upon the other: the Asian investment bubble, the dot-com bubble, the housing and commodities bubble. A dollar bubble may well be in the making.

The second problem is the total lack of any effective control over the way the system now works. The rush to "free markets" has resulted in the removal of nearly all controls formerly used by governments to regulate finances: exchange controls, limits to the capital flows, lids on speculative schemes of all kinds, rules regarding investment, and so on. World finance has become like a ship with no internal bulkheads: a breach in one place immediately floods the whole vessel, sinking it.

It is extremely unlikely that the system can be "fixed," because it is profoundly flawed to begin with and carries within itself the seeds of its own collapse. What can be done is to moderate the damage by reinstating the formerly existing "financial bulkheads": limitations on the movement of capital, exchange controls, taxes and penalties on speculative activities, investment regulations and the like. These will provide some degree of temporary stability, which can then be used to develop an alternate system. This leads to the second question listed above: is the system worth saving?

The answer is provided by the change in economic paradigms that underlies the current situation. The paradigm of growth, which is based on the abundance of cheap energy, is gradually giving way to the new paradigm of efficiency. Growth is fed by consumption, which in turn is fueled by credit. Efficiency on the contrary requires investment, the base of which is savings. The current financial system is geared to maximizing credit, which, in the current downturn, the financial authorities are desperately trying to jump start. But the era of consumption cum credit is ending, which in the end makes the current policies of "rescues" and bail-outs an exercise in futility.

These policies might well, in fact, make the situation worse. They assume that the deployment of vast sums of money provides "liquidity," which in turn should lead to the loosening and growth of credit. But the huge scale of lending and spending has another consequence: it continues to feed "the beast," the huge, anonymous and unanchored mass of money sloshing around the system in search of opportunities for profitable speculation, thus guaranteeing the arising of the next bubble, to be followed by the inevitable implosion and the resulting drop in consumption.

A complete shift in priorities thus needs to take place: the U.S. financial system must be decoupled from the global one, and at the same time reconfigured to favor investment over consumption. This brings us back to the currency and the budget.

DOLLAR DEPRECIATION—CAUSES AND REMEDIES

The problem the U.S. faces is that it is dependent on imports for about two thirds of the oil it uses. Oil is priced in dollars. If the dollar loses value with respect to other currencies, as it did between 2000 and 2008, the country's oil bill automatically rises. In order to have a chance to resolve its energy problems the U.S. needs a stable currency for at least a decade or two.

There are two main contributing causes to the depreciation of the dollar: the U.S. trade deficit and the federal budget deficit. Over the past few years, the two added together have amounted to roughly a trillion dollars a year. The current bail-out and stimulus measures will greatly increase this sum. But since their impact is still uncertain, we will stick with the older numbers, as adding to them will only increase the urgency of palliative measures.

It is questionable whether the outflows due to imports can be compensated by inflows from abroad, as has been the case in the past. The size of the deficits has grown enormously this decade, the amount of U.S. assets that can be purchased is limited, and the value of such assets depreciates together with the dollar. It can be expected that much of the trade deficit at least will accumulate abroad, generating monetary imbalances, fueling inflation, and eroding confidence in the value of U.S. currency. The budget deficit will add to these effects, since much of it is financed from abroad as well.

It does not really matter whether one considers the budget deficit or the trade deficit to be more important because they are interrelated and many of the measures taken to reduce one will also influence the other. Since the trade deficit has been the biggest, at least until recently, we will start with it.

CHAPTER SEVENTEEN

Trade

The explosive growth of the U.S. trade deficit is fundamentally due to the phenomenon of globalization. The rush to globalize has itself been based on an economic theory that has found favor in U.S. financial and government circles over the past two decades. This concept assumes that markets are inherently efficient, and that economic activity, once freed of restrictions and impediments such as customs duties, import quotas and the like, will gravitate to the areas of greatest efficiency and lowest cost. The result of this process would then be the provision of the greatest amount of goods at the lowest possible price. Once all restrictions to trade are removed and the world is, in practice, a single market, the greatest possible improvement in the general standard of living will occur at the cost of the lowest possible amount of investment. In other words, economic efficiency will automatically be maximized for the general good.

The governments of the U.K. and the U.S., the two nations where the theory of globalization has been most eagerly embraced, have thus invested a significant amount of effort in a variety of negotiations, all aimed at lowering trade barriers, either on a bilateral or multilateral basis. For the U.S. and for the countries of the European Community as well, this has resulted in the migration of entire industrial sectors to foreign countries, the chief beneficiary of this shift being China. This transfer of industrial activity has fueled the rapid growth of the trade deficits we are now struggling with.

In the "efficient market" concept underlying globalization, trade flows in one direction eventually become compensated by reciprocal flows in the opposite direction, resulting in general balance. This could possibly be true in a homogenous world where economic rationality prevails and the playing field is perfectly level. The field, however, is never level, and economic policies are always skewed by political factors. Thus, while some states will allow "free trade" to proceed according to the theory, others will practice highly nationalistic policies, with a strategy aimed at amassing wealth in the home country at the expense of its trading partners. China in particular has deliberately set the terms of trade so as to acquire huge amounts of foreign currency, while prohibiting any controlling investment in its own industries by foreigners and thereby preventing a reciprocal flow of capital from taking place. Instead of the efficiency, harmony, and reciprocity promised by the modern theory of free trade, globalization has instead resulted in massive imbalances without either effective controls or compensatory mechanisms.

The other negative consequence of the rush to globalize, and the one more closely connected to the main subject of this book, has been the bringing forward of the impending collision between the finite nature of fossil fuel resources and the economic impulse to maximize growth. The very high growth rates achieved by China and other states by means of "one-way" trade policies have put acute strain on the world energy supply, primarily for oil but for other fuels as well. This has further exacerbated trade imbalances by increasing trade deficits for energy-importing countries and adding massively to the pre-existing glut of dollars in the oil producers' coffers. This is not a self-righting situation, contrary to what the theory says. Instead it is spiraling out of everyone's control, and is to a great extent responsible for the current financial and economic crisis.

The issues outlined above are compounded by two additional factors. The globalization process often causes economic activity to migrate to areas where production labor costs are low, but where the production process is more energy intensive. This lowers overall energy efficiency and further increases the demand for fuel, with oil demand being most affected. At the same time the globalized economy relies on the intercontinental transportation of huge amounts of raw materials and finished goods over global distances. This creates an additional demand for energy, which again is mostly extracted from petroleum.

It must be emphasized here, again, that the world is not running out of oil. Reserves are still plentiful and more are still waiting to be discovered. But we have allowed a situation to be created where oil supply cannot keep up with a demand that has been artificially inflated by wasteful policies, which the U.S. enthusiasm for globalization has unduly encouraged. There are two possible ways out of this predicament. One is to do nothing until something cracks somewhere, for instance until the rising prices of energy and/or financial imbalances force a global or regional recession deep enough to temporarily break the price spiral, as is now happening. This would be a "free market" solution with no government interference, but would quickly become politically untenable. The other option is to reverse direction and reverse the progress of globalization through the application of appropriate economic policies.

When looking at this second alternative it must be kept in mind that there are two levels at which globalization operates: trade and finance. Trade globalization involves the transfer of industrial activity from high-wage to low-wage countries, usually as a result of specific policies implemented by the governments involved. It results in the creation of large trade deficits and surpluses that are funneled into the global financial system.

Financial globalization then allows these huge amounts of capital to move nearly instantly around the planet, seeking assets to invest in and fueling speculative activity, adding to the intensity of the "boom-and-bust" cycles such as we have witnessed over the last two decades. Each level intensifies and aggravates the effects of the other. The recurrence of these cycles, with increasingly destructive impact, clearly shows that globalization as hitherto practiced is *inherently unstable* and that its progress must, therefore, be halted and in many cases reversed.

The main objection raised against such a policy is that the reversal of "free trade" and of the globalization trend will bring about a major recession, if not a repeat of the depression of the 1930s. The comparison usually made to support this view is the one between the "protectionism" espoused by the U.S. during the Great Depression and the general prosperity of the first Free Trade Era in the late 1800s and early1900s. There are two problems with this comparison. First, the "protectionist" policies were adopted *after* the Great Depression started, but did not cause it. Second, and more significant, is the fact that the Free Trade Era ended with WWI. Free trade did indeed stimulate economic growth for a few decades, but it also engendered intense competition for resources and markets between the great states and empires of the period. This meant, in practice, a competitive scramble for colonies, which under the economic system then prevalent provided cheap labor as well as raw materials, together with a captive market for manufactured products.

This competition added a global component to European wars, which hitherto had been primarily territorial conflicts within the European continent. Unlike such earlier "limited" wars, the acquisition and maintenance of colonial empire required far greater resources, including high seas navies for securing global lines of communication. Thus the first Free Trade Era saw not only intensified economic competition, but also a commensurate growth of armies and navies together with the potential for conflict. Free trade had its dark side as well: the globalization of war.

It is well worth asking whether we are not currently heading into such a competitive race with the same potentially dangerous consequences. There is a current trend in that very direction. After the general military stand-down following the end of the Cold War, the United States, the chief proponent of globalization, has started rearming, doubling its military budget within the span of a decade. Its main partner in trade, China, is following suit and raising its military expenditures by 20 or 30 percent a year according to best estimates, and building a high-seas fleet. Russia and Japan are responding in kind.

We have mentioned earlier the very real possibility that the Iraq conflict is actually about oil. If this is the case, and historical logic would support that opinion, we are then headed in the same direction as the European powers were at the beginning of the twentieth century, mainly toward intensified competition for market and key resources possibly leading to armed conflict.

Finally we already *are* in a massive recession, comparable to the Great Depression according to some. "Protectionism" cannot be blamed for this state of affairs, as the world has been moving away from it for decades. But globalization, which has been intensifying right to the initial point of the current financial collapse, certainly can. There is strong evidence that points out that globalization is far from being the economic paradise its proponents advertise it to be.

By contrast what is commonly designated as "protectionism," or a partial insulation of the national economy from outside forces and interference, is not an unmitigated evil. In fact most of the industrial development of the U.S. in the nineteenth century took place behind a high tariff wall. This barrier was erected to protect the nascent American industry from the cheaper goods produced by more industrialized countries, Great Britain being the principal one. The tariff barrier did not harm the U.S. economy, which at that time was one of the most dynamic in the world. What it did was to allow the American population, industrialists and government to grow the U.S. industrial economy on their own terms, rather than those imposed from the outside by other powers. The U.S. needed such shielding because industrialization was a major change, and the nation needed its own space in order to learn how to handle it. The tariff wall erected by successive administrations gave the country the opportunity to make the needed changes and adjustments at its own pace in its own way. The changes were thus adapted to the conditions prevailing in the U.S., which were different from those in Europe. At the same time these changes were driven by the specific abilities of the American population, as well as by the uniqueness of the American political system.

The similarity with today is that our current challenge is systemic as well. Nineteenth century Americans had to learn how to deal with industrial growth. We need to learn how to do without it, or at least without the growth that we had become accustomed to. Both challenges are of a structural nature, and both require collaboration between the economic, the political, and the social spheres. Such cooperation is impossible if, while our government is national, our economy is simply a piece of the global market. We must have economic as well as political sovereignty. For this an appropriate degree of economic protection and insulation is not detrimental but, in fact, beneficial and in all probability essential.

In this respect the greatest economic disruption occurs where the trade deficit is the largest. That is also where the outflow of currency is the most significant. Such outflows are directly proportional to the degree to which the relevant foreign governments practice policies aimed at maximizing their national wealth at the expense of our own. This point is of fundamental importance. Today's mammoth trade deficits, excluding energy, are not caused purely by market mechanisms, but are to a great extent, if not primarily, the result of deliberate *government policies* involving currency manipulation and the systematic lowering of production costs.

Thus, if tariffs must be applied in order to reduce the trade imbalances, such application should not be according to industrial sector or type of goods, but according to the *policies practiced by states* where imports originate, as well as the *size of the trade deficit* with these states. The stronger the tendency of a given country to systematically generate a trade imbalance in its favor, and the greater the trade deficit of the U.S. with that country, the higher the tariff applied to imports from that state should be.

It will, of course, be objected that the U.S. has agreements under the World Trade Organization that are designed to prevent the enactment of such tariffs. There are two answers to that objection. First, the United States is a sovereign nation and is free to pursue its own economic well-being in matters that it sees as critical, of which the issue now discussed definitely is one. Second, the U.S. dollar is, for better or worse, still a critical pillar of the world currency exchange and financial structure. A collapse or major deterioration in the dollar's value would do far more harm to the world's economy than the imposition of selective, as well as necessary, tariffs. The global economy is currently at risk of being overwhelmed by its own imbalances. Any measure aimed at restoring balance and control should be implemented, in preference to more of the "free trade" that has resulted in the current crisis. It should be obvious by now that the process of globalization is not producing the expected results, and that significant adjustments are necessary for the benefit of the world in general and of the United States in particular.

A second objection is that tariffs will raise the price of goods bought by U.S. consumers and, therefore, fuel inflation. A tariff high enough to make a dent in the trade deficit will shift the production of a number of goods and services back to the U.S., resulting in the creation of jobs and in an increase of the aggregate purchasing power. On the other hand, it is no longer axiomatic that goods from abroad are inherently cheaper, particularly when energy costs are taken into consideration. The countries to which production of many articles has been outsourced are suffering from inflation as well, in most cases from higher levels of inflation than the U.S. does. To that internal inflation must be added the "energy penalty" already referred to above: the extra cost of the lower energy efficiency in the producing country. A foreign supplier may have lower labor costs, but their operations might be far more energy-intensive than is the case in the U.S. If energy costs continue to rise, as they will as long as growth is a goal the world over, the cost advantage of cheap foreign labor will eventually disappear and even be reversed by the increased energy expenditure of producing the goods and, equally important, transporting them to the U.S.

Thus the restoration of tariff barriers should not be seen simply as "protectionism," a word which implies that such measures are meant only to protect inefficient national industries against more capable foreign competitors. In the current context the use of customs duties and other import restrictions must on the contrary be seen as the necessary regulation of a runaway global financial system, rectifying the imbalances that this system generates, and preventing it from doing further damage on top of what we are already struggling with.

Redressing our trade balance is the first part of stabilizing the value of our currency. The second is to restore our budget balance, which is both possible and necessary.

CHAPTER EIGHTEEN

The Federal Budget

The second contributor to the decline of the currency is the federal budget. The federal deficit for fiscal 2008 was close to half a trillion dollars. The deficits for the following years beggar the imagination. Should the current policies be carried out to completion, their principal effect will be to inflate the "mother of all financial bubbles," based on the dollar. Once that bubble bursts, the collapse of the U.S. economy, and a great part of the global one, will be the most likely outcome. We will, therefore, base our analysis on "pre-bailout" and "pre-stimulus" numbers, on the hope that reason will prevail before a catastrophic outcome becomes likely. But we must start to seriously work on this issue now, or we will end up in major trouble, ranging from a national credit downgrade to a monetary collapse, the consequences of which cannot be clearly predicted.

The federal deficit has two main causes. The first one can be defined as "socialist mission creep." This is the tendency to look on the government as economic manager and provider, a tendency inherent in the hybridization of the capitalist and socialist political philosophies. Even though we are in principle a capitalist country, we have at the same time become used to counting on the state for the provision of a huge variety of goods and services, from safety nets of many kinds to an extraordinary variety of subsidies, make-work projects and "bridge-to-nowhere" funding earmarks. The result is a government apparatus that is becoming nearly as difficult to steer and manage as the old Soviet system. An additional source of inefficiency is that, while the Soviet system had, at least in theory, a clear command hierarchy as well as stated missions and tasks for the myriad offices and departments that made up the state structure, our government structure has by contrast been built on an "ad hoc" basis. Thus various government functions and obligations, originally inspired by a socialist-leaning political outlook, over time became not only tolerated but in the end considered as right, proper, and totally legitimate. The result of this governmental "mission creep" is that responsibilities are often poorly defined, vaguely circumscribed and split between a number of different departments.

Fixing this state of affairs will not be accomplished by legislation. It requires a national consensus, something like what was attempted with the New Deal or the Republicans' 1994 Contract with America. Such a consensus is necessary in order to re-define what the appropriate functions of government are. Without such a fundamental consensus very little will be done save for idle talk and campaign promises about how "Washington is broken and must be fixed," but without any real and working solution. Yet all the talk about the inadequacy, complexity, and ever growing cost of the federal government points out that a growing problem exists, and this problem will have to be faced in the near future. We might even be well advised to take a look at the Soviet/Russian experience of the last three decades. Hopefully nothing as traumatic as the collapse of the Soviet government will happen here, but if we do not address the issue a similar series of events will be in store for us as well.

The second cause of our government bloat is of more recent vintage; it is what could be called "the superpower complex." At the end of the Cold War, with the Soviet Union broken into pieces and Russia barely functioning as a state, a conceit arose according to which the U.S. was, for now and for the foreseeable future, the "world's sole superpower." With its main adversary reduced to impotence while its own power was intact, the U.S. now stood so much taller than any other nation that it could with complete impunity work its will on the rest of the world. This concept was not conducive to realizing, much less admitting, that we nevertheless still had problems and limitations. The logical consequence of that attitude was drift: since in our own eyes we had no problems, there was no need to exert ourselves to solve them.

Nevertheless, the nation still had internal issues to deal with and one of these was, and still is, the fact that we are living beyond our means as a nation, trusting in the continued supremacy of our currency even while we allow our financial and industrial position to deteriorate. Winning the Cold War was a glorious moment, but we cannot coast on that basis forever. In fact, we have already coasted much too long. If we are to look at the future with open eyes and clear vision, we better leave the "sole superpower" concept behind, and begin a sober assessment of where we really are and what we now need to do.

Both the above issues require a national self-examination and an honest national debate. This will take a while. In the meantime the budget still needs to be fixed to the best of our abilities, and as fast as possible. This means, in practice, bringing revenue and spending into balance.

1. Revenue

There are two ways to increase revenue: tariffs and taxes.

— Tariffs

Until the second half of the nineteenth century, customs duties were the main source of income for the federal government. Tariffs not only dampen the outflow of funds abroad as payment for imports, but they also increased the tax base by enhancing domestic economic activity. Trade generates revenue for the exporting country, but none for the importing state, unless imports are taxed at the border. Unless trade flows run in both directions in a relatively balanced manner, which is certainly not the case today, "free trade" leads to the *depletion of the tax base* for the importing state, while the exporting state accumulates wealth and grows its economy. Tax revenue is readily translatable into political power, since it allows for developing a country's infrastructure, building up strategic industries, and raising and maintaining armies. It is no wonder that an authoritarian state such as Communist China has chosen to exploit the U.S.

interest in free trade through the implementation of policies aimed at maximizing one-way monetary flows, thereby realizing a massive transfer of wealth, technology, and industrial activity from the United States to China.

Hence the recommendation earlier in this chapter that the U.S. should begin using tariffs with regard to states adopting such policies, so as to limit the outflow of currency and the consequent depreciation. Such a tariff, if selectively applied, would yield additional revenue on the order of $100 billion per year. Once such import duties begin to compensate for one-sided trade policies and to restore balance trade flows the revenue generated would begin to decrease and should ideally completely disappear. It would, however, be replaced by the tax revenue resulting from the transfer back to the U.S. of the economic activities concerned. The net result would be an immediate boost in government revenue, which would maintain itself over time.

A second recommended source of customs duties would be a tax on imported oil. Since it would be essentially a consumption tax, it cannot be so large as to cause economic hardship for the general population. As such, it would reduce currency outflow only indirectly, by making domestically produced fuel, whether petroleum-based or synthetic, more competitive. The tax take would still be significant. A 15 percent tax on a base price of $80/barrel would yield around $50 billion/year at current rates of oil imports. If domestic fuel production is increased and consumption reduced through policies such as recommended in previous chapters, the tax can be raised in order to maintain revenues, further discourage imports and provide an incentive for the development of alternative energy sources.

Import tariffs are beneficial on two grounds. One is the revenue they produce. The second is that once they are in place, they vastly increase a state's leverage over economic policy and trade relations. In this respect the current view of globalization, which assumes all duties to be evil and all foreign trade to be good, is essentially a complete surrender of economic sovereignty to non-representative international institutions as well as to foreign states seeking to acquire wealth and technology at the expense of their trade partners. It is no coincidence that the United States and the United Kingdom, the two states most committed to globalization, are also among those most affected by the developing global economic crisis. Surrendering to others the control over one's own economy is not the way to remain a world power.

— Taxes

The second potential source of additional revenue is taxation. The current situation of economic crisis is not favorable to general, across-the-board tax increases, the kind that bring in the largest amounts of revenue. The current economy is fragile at best, and the majority of the population is already struggling with a number of problems: excessive debt, falling asset prices, and a weak employment situation, to name the main ones. There are few avenues open to increase tax revenue.

One possibility, however, has already been mentioned earlier. In order to establish a valid set of market prices for the various forms of energy, *all* subsidies relating to that field must be eliminated immediately or at least phased out as quickly as feasible. This is necessary not only to establish a valid market pricing structure, but also to clear the deck for the development and implementation of future policy. Subsidies not only distort markets, but they also create and entrench forms of economic activity that are both inefficient and unnecessary, or else they would not need to be subsidized. The primary purpose of subsidies is political: they are aimed to protect and further the interests of specific interest groups. While this might be tolerable under "normal" circumstances it is not acceptable in an emergency situation such as the one we are now entering. As the withdrawal of energy-related subsidies will reduce government spending, it can be listed here together with revenue-raising measures.

Another interesting area for potential taxation is speculative financial activity. We have already mentioned how the global glut of money, fueled to a large degree by U.S. deficits, has generated a tendency to seek speculative returns and to the associated financial "bubbles." The best method to discourage such excessive and economically dangerous speculation is to tax it heavily enough to make it unattractive. If properly designed and implemented, such taxes would be, just like import tariffs, self-liquidating. But since at least part of the funds now used for speculation would be diverted into constructive investments, new economic activity would be generated and with it additional revenue.

There are, thus, opportunities to substantially increase government revenue without imposing an undue burden on the general population. The other approach to balancing the budget is of course to reduce spending.

2. Reductions in Spending

Reduction of government spending can occur on several levels. The first and most general category includes the current functions of government viewed in relation to their usefulness: are these functions necessary and justifiable, or are they superfluous? This discussion touches on the philosophical concept of government. From the socialist perspective, the state is the main provider for the needs of society, and its operation should touch every area of the nation's life. From the capitalist-libertarian viewpoint, government functions ought to be limited to national security and to ensuring the smooth functioning of commerce, everything else being left to the market and to private initiative. Our current government is, as already stated earlier, somewhere between those two poles: in theory it is dedicated to libertarian principles, but in fact it is also strongly influenced by socialist ideas and practices. The size and functions of government have grown as these practices have been incorporated into accepted government operations.

The larger question of whether and how our government is to be restructured is beyond the scope of this book. It is a long-term issue, requiring for its resolution a new national consensus that does not presently exist. A true restructuring would in any case require more data and time than is available before we start tackling the energy supply problem. We, thus, cannot, in the short term, count on any significant modification of government priorities when it comes to its currently accepted functions, and in particular its relation to social programs and the funds dedicated to them.

This leaves the following areas where spending can realistically be curtailed so as to arrive, in combination with the afore-mentioned increases in revenue, at something close to budget balance: first, a general improvement in operational efficiency; second, the elimination of earmarks, set-asides and other forms of "pork"; third, military expenditures.

— General Efficiency Improvement

Government waste is a perennial subject of political discourse. The "waste" within the federal government is described by some to be staggering. But then the size of the government is staggering as well. There is waste and inefficiency in every human activity. The larger it is, the more waste there will be. Our economy produces an enormous amount of waste, starting with the packaging of goods, often elaborately designed and manufactured for sales appeal, and yet thrown away as soon as the goods have been purchased. Yet no one is complaining, because apparently the customer likes its goods neatly pack-aged, and the supplier believes this helps sales. In that sense the government is not so different from a supermarket: we want it to do certain things in a specific way. Every interest group wants a government program or policy supporting its particular interests. Politicians then act as a conveyor for the wishes of the voters, and the required program is put in place. So we get what we ask for, and in this process no one asks questions relating to government efficiency, such as: Is this program really a proper government function? Does it overlap with another already existing program? Is the new program assigned to the correct government department? And so on. Voters get what they want, regardless of efficiency or even plain usefulness.

This is not to say that efficiency cannot be improved. But the greatest improvements can come only when the proper functions of government are better defined, so that redundancies, divided responsibilities, conflicting priorities, and other such sources of inefficiency and cost can be eliminated. This, however, must be left for later. In the immediate future not much can be squeezed out. If we can achieve an overall saving of $100 billion per year without reducing the level of service it would be a significant improvement.

— The Elimination of Pork

Earmarks, or government spending dedicated to specific constituencies, have drawn a lot of attention and provide good fodder for campaign speeches, even though the total expenditure they represent is relatively small. There are, however, many ongoing programs that were created in the same general spirit, in order to satisfy specific interests with no benefit to the general population or even designated groups of needy citizens. Such programs *can* be eliminated, but since they benefit very small and well defined groups they do not, in the end, represent very large sums. Nevertheless, some significant savings would result, and we have already mentioned energy subsidies in this context.

By eliminating such programs as well as dedicated earmarks, one could cut another $50 billion or so.

At the conclusion of this short analysis two important points must be made:

- New revenue combined with reasonable savings are sufficient to cover well over half of the pre-stimulus deficit, without imposing new taxes or hardship on the general population.

- It is important, while discussing budgets and government revenue, to look beyond the Left vs. Right divide. In this area, the debate is between the leftist policy of "taxing the rich" to increase "fairness" and the rightist template of reducing taxes on the upper income brackets so as to increase economic growth. Aside from being increasingly obsolete, these policies, formulated according to the old pattern of class warfare, will only be partially successful, and their effect temporary.

  Within the new economy, by contrast, the emphasis will be on type of economic activity under consideration, and on whether this activity is or is not in accordance with the national interest. If it is not, it should be taxed in direct proportion to the harm it can cause.

This being said, the above review is still incomplete. A major budget component, defense, has been deliberately left aside. This is because expenditures related to defense and national security cannot be looked at from a financial point of view only. One must also take into consideration how national security is understood and what it entails. This will be the subject of the next chapter.

PART

5

NATIONAL SECURITY

Over the last decade and a half our national security policy has been clouded by two misleading concepts: the "sole superpower" conceit and the blanket "war on terror" theory. Both need to be replaced by rational analysis of threats and the elaboration of realistic strategies.

CHAPTER NINETEEN

How Big a Defense Budget Do We Need?

The Cold War ended in 1991 with the collapse of the Soviet Union. For a decade after that the U.S. defense budget steadily declined, as can be expected when a major enemy has not only been defeated but has almost ceased to exist. Then starting in 2001, the military budget began to grow again, and is still increasing. In order to understand this trend, it is worthwhile to review how we got where we are; in other words, to review the historical background.

U.S. DEFENSE IN HISTORY

For most of its history, the United States has relied for its defense on a small core of professional soldiers, expanded in case of conflict through the use of mass conscription. This formula has served the country remarkably well. While the professional core ensured the maintenance of the necessary skills and discipline, the rapid enlargement of the armed forces during war drew in recruits with open minds and a "can do" attitude. Such recruits were practically oriented and more interested in "getting the job done" than in implementing pre-existing military tradition or procedures. In other words, the massive wartime manpower influx into the armed forces protected the U.S. military from the general tendency of professional soldiers to fight the *previous* war whenever a new conflict appeared on the horizon. For the U.S. each war was, to a great extent, a new one. The military won it by exploiting the national genius for improvisation, initiative, and practical problem-solving, all within the framework of traditional military discipline.

The victorious armies of WWII were rapidly demobilized according to this historical pattern. However, the Cold War had created a new situation. The expansionist policy of international communism, headquartered in the Soviet Union, demanded a reciprocal and enduring effort of containment, gradually bringing about a phenomenon hitherto unknown in the U.S.: a large standing army in a time of non-belligerence. The consequence was that U.S. military doctrine gradually became fixed on a single objective: a conventional, and possibly nuclear, conflict with Soviet forces, waged by massive armies similar to those of WWII, but technically much more advanced.

It was this armed force that was sent in 1965 to Vietnam to face a very different enemy. The Communist insurgents had learned, in the earlier campaign waged against the French military, how to defeat an adversary that was technically and materially superior. They applied the same tactics against U.S. forces, forcing a long, drawn-out fight where there was no clear path to victory. The pursuit of an apparently aimless conflict wore down the bond between the citizen soldier and the professional military. For both political and military reasons conscription was abandoned, and the U.S. military eventually became a fully professional force.

This force continued to be shaped by the confrontation with the Soviet Union for another decade and a half after the exit from Vietnam. The mission was still the same: to fight a large-scale war against a conventional adversary, on the model of WWII. But military technology had made enormous strides, and was now a main driver of military doctrine. When the enemy suddenly collapsed in 1991 the U.S. military was a hugely powerful force, with weapons that dwarfed the capabilities of any potential enemy. This force was effective against a conventional army, as demonstrated in the Gulf War of 1991. But it was also extremely expensive, and as technology continued to progress it became more expensive still.

In the meantime the Soviet collapse created what can be termed "the superpower trap": U.S. military superiority over any existing adversary was now so pronounced that it gave the U.S. free rein to intervene militarily everywhere in the world, in any way it wished, without other nations daring to resist. But the force continued to be an extension of the anti-soviet model: extremely technical, geared for a large war on a global battlefield, and relying on weapons of great accuracy, tremendous destructive power, and very high cost. There was an important piece missing: the weapons were there, but what of the target?

Saddam Hussein's 1991 army, essentially destroyed in the first Gulf War, was the last Soviet-equipped and Soviet-trained force the U.S. faced. Beyond that there is no longer a clear definition of the enemy. War with ex-Soviet Russia is very unlikely as there is nothing worth fighting over. Possible war with a rising China is in a distant and very hypothetical future. In the meantime we continue to develop weapons systems and build up our forces on a model that may no longer be valid.

While we pursue the concept of a highly sophisticated technical war, potential enemies are searching for our system's weakness. This is not a new development, for the material superiority of western armies goes back to the era of colonial empires, and the peoples subject to colonial powers needed a way to counter it even then. They picked up on the concept of guerilla war, which in the early nineteenth century was used with great effectiveness against Napoleon's armies in Spain and Russia. Guerilla warfare, practiced in all of the colonial "wars of national liberation," has matured into what is now called *asymmetrical warfare.*

The core of the concept of asymmetrical war is simple: when faced with an enemy disposing of an overwhelming advantage, develop tactics that will inflict the maximum cost on the adversary at the least cost to oneself. This type of warfare does not seek to defeat the enemy in the conventional sense, but to discourage him, wear him down, and raise the cost of the conflict to the level that the adversary can no longer afford. It does not require large forces or a large weapons inventory. In fact small forces and autonomous units are preferred while improvised or readily available weapons have proven to be most effective. Also superfluous is a large conventional command structure: in asymmetrical warfare surprise, initiative and flexibility are more important, and the military command structure is reduced to the absolute minimum.

In this new type of warfare operations are easily dialed up and down, forces shifted from one theatre to another, and tactics readily modified. The adversary on the other hand, with its much heavier investment in technology, technical systems, and structure and logistics, will take considerably more time to react and adapt.

The development and increased use of asymmetric warfare puts in question the huge rise in U.S. military expenditures in the first decade of the twenty-first century. This is not an idle question. Despite these expenditures, the continued acquisition of highly sophisticated technologies, and a truly worldwide reach, the U.S. military has, years after the terrorist attacks of September 2001, still failed to reach the single most important symbolic objective of the "war on terror": the killing or capture of Osama bin Laden and of the core Al Qaeda leadership.

In addition, none of the conflicts started under the "war on terror" umbrella is anywhere near a resolution. By contrast WWII was basically won three years after the attack on Pearl Harbor, and another nine months later it was formally over.

There is good reason to believe that our current approach to military power is not particularly effective, and that the expenditures associated with it are neither necessary nor even useful. Is there an alternative approach?

Alternatives to "Sole Superpower" Unilateralism

A major drawback of the "sole superpower" concept is that it requires overwhelming superiority in order to remain valid. Logically, this implies that the U.S. military must be capable of not only facing realistic threats and challenges, but that it can also defeat any *combination* of possible threats. It is essentially an open-ended concept. No level of superiority or preparedness will ever be satisfactory if a higher level can be aimed at.

139

Unlike escalation against a known enemy, as was the case in the Cold War, pursuing the goal of being the sole superpower is an escalation into the void, a military build-up for its own sake, focused on military power per se rather than on a realistically probable use of it.

Such an approach naturally favors unilateralism in foreign policy and strategy, disregarding the benefits to be derived from alliances, treaties and other types of understanding with foreign governments. The primary benefit of such arrangements is that they remove a whole range of threats, allowing the military to focus on such remaining threats as cannot be eliminated through means other than military. A secondary, but not negligible benefit is that treaties and other diplomatic arrangements are far less expensive than military intervention. They can, as well, provide positive benefits. We have already discussed how potential understanding and cooperation with Russia in the area of oil supply could be extended toward reinforcing stability in the Middle East, by facilitating a common policy toward, for example, Iran's nuclear ambitions. Unilateralism has temporarily taken that option off the table, leaving us with a difficult choice between an incomplete set of economic sanctions and a risky military alternative of questionable effectiveness.

Before expanding our military budget further we need to establish clearly which specific national interests such an escalation would support, and whether these interests cannot be furthered by nonmilitary means. Beyond that we need a realistic assessment of the threats we might be facing in the coming decade beyond those assumed by the "war on terror" theory. Also needed is a review of whether our current and planned force structure would be effective in meeting such identified threats. In other words, we need to put aside both the legacy of the Cold War and the questionable assumptions of the "war on terror," and then develop a military posture that answers the real challenges of the future.

REDUCING EXPENDITURES

Such a process should go in parallel with the winding down of the Iraq war and other related conflicts. These wars have resulted in a significant wearing down of our inventory of weapons and equipment, to the point where units no longer have their original complement, and equipment must be rotated between them as they move in and out of the war theatre. Before the equipment is replaced "as is" it must be determined whether such replacement is necessary or even useful, or whether new concepts in force structure and armament should be introduced instead.

It is such *new* concepts, rather than the simple maintenance and replacement of what we have, that should have first priority as far as spending is concerned. There is, therefore, a window of opportunity for a substantial reduction of military expenditures in the short term. Half of this reduction would come from the winding down of war expenses, the other from reversing the military budget growth of the last six or seven years. The cumulative savings as the withdrawal from Iraq and other conflict theatres is completed would be in the range of $300 billion per year.

A Fundamental Choice in Foreign Policy

The founders of the United States did not mean it to be an imperial power. Even if they had foreseen how large and powerful the country would eventually become, they would not have wanted it to be a potential successor to the British Empire, against which they fought the War of Independence. At various stages since its birth the nation has been tempted by the imperial dream. Occasionally it succumbed to the temptation, but each time it drew back, having realized where that road ultimately led away from the ideals and principles upon which the nation is founded. The choice made by the Founders is that the United States be an inspiration and an example rather than a master. If the U.S. is to lead, it is to do so by example rather than by force.

The global reach of American power after WWII was neither willed nor planned. It was the result of circumstances thrust upon the nation. These circumstances, and the conflicts that the U.S. had to fight and win, had their roots in the ambitions of the totalitarian states that arose in the twentieth century. Even after its complete victory in 1945 the U.S. proceeded to promptly demobilize its huge wartime forces, until it was faced with the aggressive advance of world communism in both Eastern Europe and China. The Soviet Union is now dead, and while China remains somewhat of a question mark, the need for a continued U.S. global hegemony can be put in question, particularly if such hegemony is supported by military power and demands submission rather than acquiescence and cooperation.

The "sole superpower" concept, backed by the "war on terror" theory, looks like another version of the periodically recurring imperial temptation. There are alternatives, and we would be well advised to give them serious consideration.

Before going further on the subject of defense and security, however, we will briefly return to the budget issue, which is the subject of the previous chapter.

ADDING IT UP

The additions to revenue and the reductions in spending that have been suggested in the last two chapters would roughly equal the average annual budget deficit the U.S. would be looking at through the next few years, under normal circumstances. These measures will put the federal accounts into balance without a significant reduction of government services. It needs to be mentioned again that this balance does not include the recent "bail-outs" and "stimulus packages" and overreaching budgets, which will sooner rather than later have to be abandoned for being both unsustainable and unnecessary.

As for the trade deficit, the measures recommended would leave only oil and other energy imports as significant items. These in turn would be addressed by the energy-related policies recommended earlier.

The entire package will, therefore, address the two main causes of dollar depreciation. In addition, the domestic investments resulting from the recommended policies would be creating new assets, which in turn would expand the potential for foreign investment. The inflow and outflow of currency would, thus, be tending toward balance.

Before the defense budget can be set at the proper level, however, we must have dealt with the issue that has been used this decade to justify ever higher military expenditures: the so-called "war on terror," and the problem of terrorism. This will be done in the following chapter.

CHAPTER TWENTY

Dealing with Terrorism

National security, particularly against terrorism, has been the main priority of the U.S. government during this decade. It is our conviction that the dependence of the nation on foreign energy sources actually is, in the medium and long term, a far greater threat. But terrorism is spectacularly murderous, and it is far easier to mobilize the electorate against villainous assassins than it is to persuade the voters of the seriousness of the energy issue. In order to reset the order of national priorities, it is necessary to review the terrorist threat and assign it to its proper place and importance.

Since 2001, the United States has been engaged in the "War on Terror." As presented by its supporters, this struggle pits the United States and other western democracies against a radical Muslim conspiracy. The goal of this alleged Islamic movement would be nothing less than the destruction of the western way of life and its replacement by a doctrinaire and aggressive brand of Islam. Al Qaeda, the terrorist group that perpetrated the September 2001 attacks on the World Trade Center and the Pentagon, is the visible leader of this movement, which is assumed to have many sympathizers, supporters, and participating organizations. This radical Islamist coalition has been pictured as presenting a mortal danger to the West in general and the U.S. in particular. Its defeat can only be achieved by a single-minded and intense campaign stretching over many years and demanding the utmost dedication.

The general umbrella of the "war on terror" covers a number of specific U.S. policies, among which are: the effort to contain Iran and frustrate its nuclear ambitions; the isolation of Syria; the wars in Afghanistan, Iraq

144

and Somalia; the establishment of a separate U.S. command in Africa; the pressure on Pakistan to police its tribal territories; and a number of covert operations spanning most of the globe.

Because of the transnational character of the alleged Islamist conspiracy, the "war on terror" also serves as one of the justifications for the continued U.S. military build-up and the establishment of a growing number of U.S. military bases throughout the globe. The life-or-death character of this alleged struggle against "radical Islam" has been successfully used as a political argument in two U.S. elections. It has also been a powerful motivator behind the passage of a number of security-related laws and the associated regulations.

In any war or conflict the first requirement for victory is to know and understand the enemy. It is only on the basis of such understanding that the most effective strategy and tactics can be developed and implemented. Seen in that light, the phrase "war on terror" is a misnomer. Terror is a concept, a tactic or a method. It is something the enemy might use or implement, but in itself terror is not the enemy. The "war on terror" concept defines the enemy as either terrorists or radical Islamists. Again this creates confusion because terrorism is not a specifically Muslim trait. It is a recurring phenomenon that has manifested itself throughout the globe since the middle of the nineteenth century.

Modern terrorism was developed in Europe in the 1800s, in connection with the ideologies of anarchism and revolutionary socialism. It has been practiced in a great variety of locations and by many different groups, only some of which have a connection with Islam. As far as the Middle East is concerned, many "Islamist" organizations have no connection to terrorism. In fact, original practitioners of terrorism in the Middle East were *secular* organizations, such as Yassir Arafat's Fatah or George Habash's PLFP. Terrorism by militant religious Muslims is by contrast a relatively recent phenomenon.

In order to understand the challenge we face with respect to terrorism, we must, therefore, ask and answer, three questions: What is terrorism? What is Islamism, radical or otherwise? To the degree terrorism and Islamism overlap, what danger do they represent and how are they to be fought?

THE NATURE OF TERRORISM

Terrorism can be loosely defined as the use of essentially military tactics against civilians and other nonmilitary targets. The general goal of terrorists is to use such attacks, or the threat of them, as a means to force the adversary to accede to the terrorist organization's demands.

Modern terrorism originated approximately a century and a half ago and its history includes many groups, organizations, parties, factions, and individuals, as well as a variety of core ideologies used by such groups to justify their actions. Such groups include, among others: the Russian anarchists and social revolutionaries before the 1917 communist revolution, including large organizations, small autonomous cells or even lone individuals; nationalist groups in Serbia, Israel and Ireland; a number of Palestinian entities, first secular and later Islamist; radical Marxist entities which arose, mainly in the 1960s, in Western Europe, South America, Japan, and the United States; ethnic separatist groups in Spain and France; a cult in Japan; and finally the "Jihadist" entities in Muslim countries, operating in the broad Middle East including Egypt, Afghanistan, Iraq, Pakistan, Algeria, and Morocco, as well as countries further east.

The first notable characteristic of these groups is their diversity. They appear in both the East and the West, as well as anywhere in between. Some are, at least nominally, religious; some secular; and some aggressively hostile to religion. They also pursue a great variety of causes: national or ethnic independence; social change; a shift from autocracy, or dictatorship, to democracy; national and/or global revolution; religious reform; even animal rights. There is no visible and predominant connection between terrorism as a concept or tactic and any specific culture, ethnicity, or state.

There is, however, a striking similarity between the genesis and evolution of terrorist groups, cutting across boundaries and cultural differences. All terrorist groups originate and develop in a remarkably similar way. This development includes the following phases:

- **First Phase: Participation in a Wider Political Movement**

  Every terrorist group begins its life cycle as part of a broad-based political movement. The goal of the movement may be purely political, such as the replacement of autocracy or dictatorship by a representative system. It can be economic, such as the shift from economic oligarchy or capitalism to a socialist structure. The goal can also be nationalist or ethnic, seeking independence or autonomy for a particular ethnic or cultural group within a larger political structure, such as an empire or a nation ruled by a different ethnicity. Finally, the goal can be religious, seeking the incorporation of specific religious or moral rules into the prevailing political or legal system. Often it is a combination of some or all of the above, but there generally is a primary driver.

  In any broad-based movement there will be a range of attitudes, factions and projected timelines with respect to the achievement of the collective goal, spanning the spectrum from moderate and patient to radical and in-a-hurry. Terrorist groups will always originate at the radical end, as can be expected.

- **Second Phase: Factional Split**

  Political movements normally seek to achieve their ends through legal and peaceful means: the creation of a party, petitions, marches and demonstrations and strikes. The more moderate factions also seek to develop connections with favorably inclined members of the existing power structure. In many cases the above methods will be successful in achieving the desired end, but there is always a certain amount of resistance that slows progress down, and often brings it to a halt. In many cases, the government's resistance to change turns to repression, with forcible break-up of demonstrations, intimidation and arrests of leaders, vote tampering, proclamation of martial law, and passage of restrictive laws and regulations aimed at weakening the movement and slowing down its growth.

Lack of progress and repressive measures inevitably give rise to frustration and discord within the target political movement. In general the majority of members opt for continuing the work, even if success is limited and little progress appears to be made. The radical faction(s), on the other hand, will tend to see and denounce such compromise and temporization as surrender and failure. If unity cannot be maintained and compromise goals agreed to by all, the radicals will at some point decide that the existing political order presents an insurmountable obstacle and must be overthrown before the movement's goal can be achieved. They shift from an evolutionary to a revolutionary approach, with armed struggle as a final resort. This then causes an ideological split within the broader movement, usually ending with the radical fraction being expelled or splitting off on its own, becoming an autonomous entity with its own name, agenda, and leadership structure.

- **Third Phase: Terror**

Once freed from the restraints imposed by the larger movement, the radical faction eventually embraces "armed struggle" and begins a campaign of terrorist action against a number of targets, such as high government officials, members of the security services, critical industries, or simply the population at large. The aim is to weaken the government, cause it to adopt unpopular repressive measures, and create a chaotic situation favorable to the desired revolution.

Almost invariably, the results are counterproductive, even if a number of the listed targets are actually hit. The attitude of the government hardens, security is increased, and the radical group, now a "terrorist" outfit, is forced to go underground. Secrecy becomes paramount, and in order to survive the group must adopt special measures to avoid detection and infiltration by government agents and informers. It, therefore, must limit contacts with the outside and set up several layers of security between the core leaders and operatives and their wider network of sympathizers and supporters.

This physical isolation from the general population and even from sympathizers has two consequences: first, recruitment becomes both difficult and dangerous, since the authorities will do everything possible to penetrate the terrorist network. The addition of new members and operatives will be extremely slow, particularly at the leadership level. The same difficulties will be present in the areas of funding and communications, for which secure and circuitous channels, insulated against detection and eavesdropping, must be established. This need for secrecy and careful vetting of the membership will inevitably reduce the size of the terrorist group and make significant growth in numbers nearly impossible. It will also crimp its access to material resources.

The second consequence is that once the group has gone underground, the leadership will lose any direct contact with the general political scene, and any exchange of ideas with others than members of the group will be cut off. The core ideology of the terrorist entity will then tend to harden into a permanent form, within which the "armed struggle" will take ever increased importance. Terrorist violence, once initiated, becomes an end in itself, while the former and broader political goal is gradually reduced to a prop. While the group may long remain a threat and perform a number of "successful," meaning deadly, attacks, it has now reached an ideological dead end. This eventually deprives it of any meaningful political influence. The revulsion caused by terrorist mayhem further isolates the group from the general population, including in most cases the supporters of its original political goals.

- **Final Phase: Attrition and Death**

Once the group has locked itself into the terrorist template, it can no longer grow. As the government authorities have vastly greater resources at their disposal they will eventually succeed in grinding it down through sheer attrition. Experienced operatives will be lost one by one, and eventually the core leadership will be unmasked, located, and either destroyed or taken into custody. This may take a decade or more, but history has shown it to be inevitable. Governments have been damaged by terrorist operations, but never destroyed. It is possible in the extreme that a government will be so weak, divided, and riddled with corruption and incompetence that it will be incapable of eliminating a terrorist threat, which will continue to fester in the chaotic situation that government weaknesses allows to develop. But even then terrorism will not bring about a change or a revolution. Only determined, public, and organized political action will achieve such an end. No terrorist group in history has ever succeeded in achieving meaningful change.

On the other hand, no terrorist organization has survived a determined counterterrorist effort by governmental security agencies. All were ultimately destroyed, disbanded on their own, or driven to take refuge abroad.

There are several conclusions to be drawn from the above:

— Terrorist organizations are by necessity small, secretive, and limited in terms of resources and manpower. They include, however, skilled and dedicated leaders and operatives. The forces needed to fight them need not be numerous, but must be capable, highly motivated, and expert in stealth and undercover operations

— Antiterrorist operations require first-class intelligence, much of which is specific to the area and population within which the terrorist network operates. Operations directed against terrorist groups require extensive cooperation with local authorities and residents. In cases of transnational networks, such as Al Qaeda, close cooperation with other national security agencies will also be essential.

— Terrorist groups must be isolated from other militant groups that might offer them assistance, shelter, and cover. In a mixed situation, such as exists in Iraq, Afghanistan and Somalia, it is important to distinguish between such groups according to their goals and objectives, some of which may be fully legitimate. Putting all adversaries, regardless of their respective objectives and methods, under the designation "terrorist" only allows genuine terrorists to gain allies and support.

— All terrorist groups start with a cause, which can be legitimate and have the support of the general population. Such a cause often includes a drive for normal human rights as well as for the redress of genuine grievances. While any particular terrorist group can be eliminated, if such grievances are not addressed, new groups will continue to arise through the radicalization process outlined above. In the end, the elimination of terrorism, as such, requires the establishment of a government that is accepted and supported by the people and that basically answers to the legitimate aspirations and needs of the population.

The above points, which are germane to all terrorist entities without exception, point to the most effective methods to combat terrorism and prevent it from arising.

EFFECTIVE COUNTERTERRORISM

The first and most important lesson to be drawn from the above analysis is that large-scale military operations, such as the U.S. has carried out in the Middle East under the banner of the "war on terror," are the least efficient means of controlling and eventually uprooting terrorism. Such operations have the following drawbacks:

— They do not address the fundamental grievances or aspirations of the population, and may in fact create additional problems, such as non-combatant casualties, collateral damage, material destruction, insecurity, displacement of populations, and so on. To this must be added the general dislike of any population for military occupation by foreign forces.

— They generate opposition, both political and armed, from a variety of opponents. This allows the genuine terrorists to operate freely within the chaos and disorder created by an invasion, to make allies, to enlist recruits and develop support networks.

— They generally depress and/or disrupt the economy and lead to large-scale emigration, caused by the lack of personal safety and by economic insecurity. Such emigration primarily involves the professional classes, resulting in the degradation of government functions, social services and economic management.

— They are extremely expensive. This includes lives lost, both among the civilian population and the armed forces. Added to that is the expense of keeping large numbers of troops in a hostile environment as well as the attrition and loss of costly equipment.

These drawbacks not only make the cost effectiveness of such operations very low, but might actually worsen the situation rather than improve it, since opposition to military occupation will eventually create more enemies than are eliminated by it. This creates a self-perpetuating and disorderly situation within which terrorists can not only operate but actually grow in strength and numbers.

If the mission really is the elimination of terrorist groups and of the causes of terrorism, then any conventional military intervention and/or occupation should be terminated as quickly as practically feasible, and replaced by the implementation of a comprehensive strategy specifically adapted to the terrorist threat. This strategy *must* address not only the immediate terrorist activities but also *the underlying causes and circumstances* from which the terrorist threat emerged. The necessary components of such a strategy are the following:

A. Identify and Address the Underlying Grievance(s)

Terrorist entities are violent offshoots of broader movements, the great majority of which are legitimate and have been established to address real problems and needs. Such issues and grievances can be of many kinds: ethnic, cultural, economic, or political. As long as the fundamental issues are not addressed, there will be discontent among the population, the radicalization process outlined above will proceed, and terrorist groups will not only appear, but also benefit from a degree of general sympathy and support. As a consequence, information on the identity, location and moves of terrorist cells will not be forthcoming, resulting in poor or incomplete intelligence. In such a situation terrorists will not, in the eyes of the population, be seen as criminals as long as their targets are perceived as being enemies or oppressors of this population.

The power that is targeting the terrorists must develop a broad strategy that will answer the needs and aspirations of the population. In the Middle East, where many states have artificial borders left over from past colonial empires, one primary popular aspiration is ethnic autonomy and self-determination. Economic development will generally be the second most important priority. A third would be the respect of local power structures and hierarchies, as opposed to the imposition of a distant and often corrupt central government, be it nominally democratic or not.

B. Develop an Action Plan and Allocate Resources

Once the key needs and grievances have been identified, a plan to remedy them must be put together, and the needed resources must be allocated accordingly. Such resources need not be huge, because the goal is not to totally transform the area or region, a process often referred to as "nation building." For the purpose of counter-terrorism it is sufficient to gain the goodwill of the population. As long as the plans for attaining this are credible and their goals genuine, this population will be ready to make a major contribution to the anti-terrorist effort. This will usually take the form of providing a security force or militia to protect their territory, together with the support and intelligence that makes such a force effective. Most areas in the broad Middle East have some kind of local security force or militia, which often arises spontaneously and can be very effective within its own area, since it intimately knows the people and the terrain. Such militias must not be considered as rogue groups, as is still the case in Iraq and other locations.

In such areas U.S. policy has generally aimed at disarming the "militias" and creating a "national" security force. This policy ignores the fact that in most locations such a central force has, often for generations or even centuries, been seen as foreign, oppressive, and generally corrupt, which is why local militias exist in the first place.

Once a reasonable and realistic plan has been elaborated, it must be communicated and clearly explained to the population, which should have been consulted beforehand in any case. The goals, timelines and requirements should be clear from the start if the desired support and cooperation are to be obtained.

▪ Implementation

It is to be noted that the recommended strategy *does not begin with military operations*. If some reliable information concerning the target terrorist group becomes available, it could lead to a strike, but the main goal is to *first and foremost gain the support of the population*. Without this cooperation a terrorist entity cannot be destroyed or neutralized. For this reason the development plan referred to above must be implemented first in the areas that are most secure, allowing progress that is both quick and relatively inexpensive. These areas can then be used as showcases to bring other districts to the table. Starting in a secure spot with a program that benefits the population sends around the message that supporting or allowing terrorism does not pay, while cooperation with the anti-terrorist entity is beneficial.

It is at this point only that a counter-terrorist strike force be introduced, because it will be seen as a protection force rather than an occupation army. In all likelihood the terrorist leaders would have learned of the development program and concluded that it represents a danger to their support among the population, support which is often obtained by terror or other forms of coercion. They will, therefore, attempt to disrupt the development program, presenting the locals with a choice: either keep the program going and protect it jointly with our own forces, or see it being destroyed and its benefits disappear. If we have done our advance work well, the first choice will be the most likely. It will put the onus of the first hostilities on the terrorist organization, creating a dilemma for their supporters and hostility among those who risk becoming the targets of terrorist strikes.

If the strategy is well planned and executed, our own forces need not be large. The enemy will not be numerous, and general security as well as intelligence can and will be provided by the local population. The total cost, including the development program, will be far less than the conventional military occupations that have been the standard practice within the framework of the "war on terror."

C. Get the Job Done and Get Out

The proposed strategy is a minimalist one. It aims at one objective: the elimination of terrorist groups and networks. It begins with determining the basic needs and aspirations of the host population; it then continues with the development of a plan that will satisfy these needs and gain the population's support; this is followed by the implementation of the plan, starting in the *most favorable location* so as to rapidly create an example of success, and using local resources to the maximum; finally, as the plan begins to succeed, forcing the terrorist entity to fight at a disadvantage or lose their credibility and their support. Once this has been achieved and the enemy is destroyed or neutralized, we must make clear our intention to leave as soon as possible, and act on it. If eliminating terrorism is truly the objective, there is no need to modify existing social or political structures, promote our style of democracy, sign treaties or build bases. Fighting terrorism will be most effective, and least costly, when this objective is clearly separated from global power plays and imperial designs.

THE ISRAELI EXAMPLE

An excellent example of both the success and the failure of counterterrorism efforts is provided by the Israeli operations against the Palestinian entities that use terror as a weapon. These operations have a long and well-known history and can, therefore, reliably serve as a reference on what does and does not work.

The initial "grievance" in this case is territorial in nature: the Palestinians, which have left what is now Israeli territory as a result of successive wars have been prevented by Israel from returning. At the same time they were not allowed to settle in neighboring states by the respective governments, and were kept indefinitely in refugee camps. Since the 1948 war, which created the problem, the stateless Palestinian population has grown several times over, a development which makes such return or resettlement difficult, if not impossible. The Palestinian "refugees" thus remain the only major displaced population, since WWII, that has not been successfully resettled in a new national home.

If such a resettlement could somehow be contrived, it would essentially solve the problem. Unfortunately very little has been attempted in that direction, mostly for political reasons, and the situation continues to fester, generating terrorist entities in the process. Such terrorist activities are directed primarily against Israeli targets.

The Israeli counter-terrorism operations against Palestinian groups, who are now in their second generation, have been both a tactical success and a strategic failure. The Israelis have learned, through their experiences in the Gaza Strip and in Lebanon, that the occupation of hostile territory by conventional forces is both excessively costly and fundamentally ineffective. At the same time they have demonstrated the effectiveness of targeted strikes based on timely and accurate intelligence. Such strikes, aimed essentially at enemy leaders, safe houses and support facilities, have decimated the militant leadership and kept it from carrying out coordinated and effective operations.

But the continuing tactical successes contrast with a parallel strategic failure. Because both sides essentially claim the same territory as their own and the Israelis have effective control of it, the basic grievance of the Palestinians cannot be addressed under current circumstances: two different populations cannot exercise sovereignty over the same land. And after 60 years of opposition, several wars and more than a generation of terrorism and counterterrorism, it has become practically impossible for the two parties to agree on anything of substance. The Israeli security forces can, and do, destroy one terrorist cell after another, but new ones constantly appear. The cycle is never broken, and any relaxation of Israeli vigilance allows the terrorist danger to reappear and grow again.

A similar situation exists in other regions where comparable problems are present: the Kurdish areas of Turkey and Iran, the tribal territories of Pakistan, southeastern Afghanistan, and other such. In each of these areas, endemic terrorism can be contained by the appropriate tactics, but its elimination requires policies addressing the aspirations of the population. Until such policies are in place, the problem will continue to smolder, flaring up every time circumstances allow.

ISLAMISM

Here an additional word needs to be said about Islamism. The alleged enemy in the "war on terror" is usually identified as "militant Islam" or "radical Islam." The underlying assumption is that there is within Islam a new and aggressive trend aiming at mobilizing Muslims for a global war against the West, and particularly against the United States. This concept makes terrorism connected with Islam appear to be a new and more virulent form of the use of terror, an interpretation that can lead to costly strategic mistakes.

The first relevant fact regarding Islam in this context is that for the last three centuries the Muslim religion has been anything but aggressive. The last attempt at military conquest by an Islamic state was the siege of Vienna carried out by the Ottoman Turks in 1683. Vienna was not taken, and the Ottoman army was soundly defeated before its walls by a relief force led by the king of Poland. Since that time the expansion of Islam has proceeded exclusively through peaceful proselytism, which was quite successful in Southeast Asia. By contrast, the main globally aggressive powers for the last five centuries were the European states, which over that same period conquered and colonized the Americas, the whole of Africa, and a good part of Asia. During that period most of the Middle East was at one time or another occupied by the British, French, and Italians. Many of these conquests resulted in the long-term occupation of the territories concerned, reducing them to the status of colonies and their inhabitants to the level of second-class citizens under European rule.

In the Middle East and other Muslim lands, this trend toward European expansion and domination was reversed in the late nineteenth and early twentieth centuries, when Muslim elites began to adopt western technology and methods of administration. As a result, increasing numbers of the local upper class received a western-style education, which trended more and more toward secularism. Comparing the "advanced" state of European society with the "backward" conditions prevailing in Muslim lands, these reformers often concluded that Islam was a main contributor to their populations' backwardness. They reasoned that if the hold of Islam over the population was broken and replaced by modern secular thinking, their countries would rapidly catch up with the West in terms of technology, wealth and power.

This secularizing trend was dominant among the elites of Muslim lands through the first half of the twentieth century. It was often mingled with sympathy and admiration for Nazi Germany and Soviet Russia, which were seen as "underdogs" struggling to gain power and recognition, just as the Muslim states and other colonized peoples were attempting to do. Various forms of secular socialism became the preferred political doctrine of Middle Eastern reformers, even while Islam was being pushed into the background and regarded as a retrograde influence.

This drive toward a secular and state-driven transformation of Muslim society had two main consequences. First, it suffered from all the drawbacks of socialism: an arrogant and oppressive state, low economic efficiency, the absence of outlets for private initiative and contempt for traditional culture. The subject populations were told that they must westernize themselves, but they saw no benefits in this effort. In the Middle East specifically, the repeated defeats of Arab armies at the hand of Israeli forces discredited the political leaders and, by implication, their secular socialist ideologies. The combination of authoritarian rule and simplistic socialist creed caused the gap between Middle Eastern and Western societies to grow, while the bulk of the population remained mired in poverty and stagnation.

This general failure of western-style secularism reflected on attitudes toward Islam as well. It enhanced value of the Muslim creed and culture in the eyes of the common people. Against the backdrop of secular failures Islam continued to provide a foundation for daily life and a sense of collective identity, reaching back in history to a time when Muslim states had been rich in power, wealth, and intellectual achievement. The failure of socialism and the oppression perpetrated by secular regimes naturally turned the people back toward this heritage. This renewed emphasis on Islamic values then led toward attempts to organize society on the base of Muslim, rather than western, tradition and thought.

The resulting movement or trend is generally called "Islamism," and aims at re-integrating Islamic tradition, ethics, and jurisprudence into the legal system and the operations of government. Islamic political parties have generally been better organized and less corrupt than their secular counterparts, and many of them have managed to provide the population with essential services that the secular state failed to deliver. As a result, they integrate better into the life of the people and tend to generate much more enthusiastic support among the population than the secular parties built on the western model. A contributing factor is that traditional secular parties are generally subservient to authoritarian or dictatorial governments and, therefore, lack a popular power base and an agenda of their own.

Islamist parties have, from the beginning, provided a genuine challenge to the existing power structures, which are predominantly secular and generally supported by the military and internal security establishments. For that reason Islamist parties have often been repressed and put down by force, particularly in Egypt, Algeria, Turkey, and Syria. This has led to a split between the moderate Islamists, who do not support armed struggle and aim at obtaining power in legitimate ways, and radicals willing to embrace violence and terrorism. According to available information, the moderate brand of Islamism has by far the greater popular support, while the support for the radical factions has been dwindling in recent years.

In summary, the above analysis brings out the following facts: first, the rise of Islamism is primarily a phenomenon internal to Muslim states, seeking domestic reforms along the lines of Muslim tradition rather than on the model of western secular humanism. As such, Islamism as a movement long antedates the "war on terror." Second, within this movement the moderate wing is by far the predominant one in terms of numbers and popular support. Even those who have been radicalized by repression and have taken up arms or bomb making focus primarily on the domestic scene and by no means constitute the international underground army that the western proponents of the "war on terror" assume to exist.

The opposition of such radical groups to western influence stems primarily from the association of western powers with repressive Middle Eastern regimes, and this opposition has, of course, grown as a result of large-scale U.S. military intervention in Afghanistan, Iraq, and elsewhere. Another negative for western states is their still recent colonial past, during which the resources of the occupied territories were appropriated by the colonial powers. The colonial template of the occupiers exploiting the occupied is, of course, easily transferable to the issue of western thirst for Middle Eastern oil.

But to the extent that Islamists and others oppose U.S. presence, their motives are primarily political and, if oil is involved, economic rather than religious. Such is certainly the case for any armed or other resistance to military occupation. Even Al Qaeda's nebulous "caliphate," a future Muslim commonwealth similar to the Islamic empire created by the early followers of the Prophet Muhammad, is essentially a political concept. Few Muslims aside from Osama bin Laden find the realization of this dream to be possible or even desirable.

Terrorism is not proper to any particular time, nation, or culture. It appears whenever legitimate popular aspirations are thwarted to such an extent that some militants or activists, for a cause that itself is perfectly valid, "fall off the cliff" into violence. The root cause stems not from their culture, ethnicity, or political orientation, but from an inability to cope with failure and the frustrations it engenders. Identifying terrorism with a particular nationality, political persuasion or religion shows a fundamental misunderstanding of its nature and origins.

OVERREACTION VERSUS VIGILANCE

The "war on terror" is, in fact, an amalgam of two different policy streams. The first is the campaign against terrorism proper, which has been waged, from the military point of view, with the proper tactics and weapons: the pursuit and acquisition of intelligence, followed by highly targeted strikes. These tactics have been responsible for our successes to date against Al Qaeda and related groups—successes that are substantial and have greatly weakened the adversary.

The second stream is a series of interventions by large conventional military forces, both our own and those of various proxies, in several locations in the Middle East. The very broad, and somewhat nebulous, justification for these large-scale operations has been: "if we don't get them they will get us." "They" in this case stands for an assumed pan-Islamic conspiracy against the West, a conspiracy which in fact simply does not exist, although a succession of military interventions, if carried out long enough, could eventually engender such a broad form of opposition.

The first part of this "war" has been successful and should be intensified along the lines sketched out earlier in this chapter. Such operations can be made even more effective through increased cooperation with other national security services facing similar threats. The second part—by far the most expensive in terms of lives and financial cost—has probably been counterproductive, meaning that aside from the massive amount of damage it has inflicted, it has probably generated more hostility and raised more enemies than it has eliminated. This has little to do with the performance of the troops, which by all measures has been outstanding. But military skill, courage, and professionalism cannot fully compensate for the drawbacks of an ill-conceived policy.

The other drawback of the "war on terror" is that it has been a huge distraction. Our foreign and domestic policies have been focused almost exclusively on terrorism and its implications, both real and imagined. Even in the 2008 election the ability to handle national security and to prevent terrorist attacks was considered a prime requirement, and so far no one has dared to say: "Well, terrorism is indeed an issue, but we have others, and some of these are more important and pressing." An objective analysis and evaluation of the terrorist threat are still absent from our political discourse.

The fight against terrorism, using correct tactics, must be continued and even intensified. But it must also be ranked among the other problems we face. The energy issue, if misunderstood and mishandled, can not only hamstring our economy and severely affect our way of life, but it can at the extreme lead to virulent international competition, including armed conflict, for access to, and control of, energy supply. Wars have been fought over gold and they could just as easily be fought over oil.

A failure to deal with our energy issues will, in the long term, have greater consequences than terrorist attacks. But beyond its strictly national impact lies the fact that energy is also a global problem. To address this issue will require not only major effort on the national level, but extensive arrangements and cooperation on the global scene. Dealing with energy, more than anything else, can truly be termed "a challenge for generations."

Answering this challenge will have a political as well as an economic dimension. The political one requires the development of a clear national consensus and a supporting political formation. The next, and last, chapter of this book will deal with this crucial issue.

6

WHERE TO NOW?

The energy situation presents America with a major challenge, but our history shows that the nation has successfully met such challenges before. What the nation has done in the past we can now do again.

CHAPTER TWENTY-ONE

A Glimpse Into the Future

The United States today is faced with a multiplicity of issues: health care, immigration, how to reign in entitlements and many others, all requiring attention. Nevertheless the energy issue is more central and fundamental than any of these, for both economic and political reasons.

This does not mean that other issues are unimportant. They are not, and the way they are addressed will have major impact on our way of life. But the energy issue affects not only the resources that will be available for their resolution, but, more importantly, the principles on which such resolution will be based.

Let us look at the issue of resources first. All government budgeting in the modern era has been based on the assumption that state revenues will grow indefinitely. If this is true, then running up the national debt and adding new entitlements to those already in place is quite reasonable. Economic growth will, in that case, guarantee ever increasing tax revenue, from which additional government programs can be financed, by further borrowing when necessary.

If on the other hand limitations on energy supply restrain or even halt economic growth, other approaches to providing essential services will have to be devised. It will no longer be possible to do so simply by increasing government budgets and voting appropriations for new programs. If the current energy shortage eventually develops into a full-blown crisis, we might not be able to afford, economically and financially, a solution to any of the other issues included in the above list.

Energy is so vital to our economy and way of life that, once its supply becomes restricted or unstable, we are in unknown territory and to a great degree all bets are off.

These are the material facts, from which we cannot escape because of our physical dependence on the energy supply. Since, however, the decisions involved in attacking the energy situation are, to a great extent, political in nature, another angle must be considered: the starting point from which a political process is initiated, and the principles that guide the development of this process.

We have touched, at the beginning of this book, on the subject of the great divide between Right and Left in modern history. We have also shown that, to a great extent, this separation between capitalism and socialism, between individualism and collectivism, between the power of the state and the untrammeled freedom of the individual had its root in the disposition of the wealth created by the Industrial Revolution. And that particular historical and economic episode is inseparably bound with the use of fossil fuels as a source of energy.

Now we have come to the initial phase of the conclusion of the Industrial Revolution episode as it has been known and lived for roughly two centuries. Its primary premise, that energy will be indefinitely available at a low price, is gradually weakening and will continue to do so. The path of endless economic growth is beginning to shut down, and a new paradigm centered on energy efficiency is taking hold.

As this transition, which will take a generation or two to be completed, now begins, it behooves us to recognize that it is not only the assumed endless abundance of physical energy that is ebbing, but that the ideologies it generated are becoming obsolete as well. They need to be abandoned, and then replaced by principles better fitted to the new situation. To briefly return to the initial chapters of this book, the two ideologies that have dominated the modern era up to now have been:

- *Socialism*, which gives the state full control of the economy and of everything the economy produces. Socialism has been

discredited by its failures in the Soviet Union and other communist states, and is no longer considered viable in its pure form.

- *Capitalism,* which minimizes the role of the state and leaves
economic control to entrepreneurs and corporations, trusting
in the wisdom of markets. In its latest globalized incarnation
unfettered capitalism has briefly triumphed after the collapse
of the Soviet Union, but the global crisis that started in 2007
has put an end to this ephemeral reign.

Neither the proponents of capitalism nor those of socialism recognize that their respective systems have any drawbacks, but see only
their positive characteristics: capitalist ideologues focus on individual
creativity, socialist ones on fairness. In fact both systems have substantial
shortcomings. Capitalism in its pure form certainly does lead to rapid
economic development, but also to worker exploitation, growing income
disparity, and environmental destruction. Socialism on the other hand
does result in a form of economic fairness, but it also fosters economic
inefficiency and the creation of a vast bureaucracy that is ultimately more
interested in its own power and privileges than in the general progress.
The bureaucratic elite is also inclined to use police state methods to
maintain and improve its position, a development that early socialist
ideologues such as Lenin saw as a necessary and legitimate characteristic
of revolution.

The supporters of either system use the negative outcomes listed
above as arguments to criticize the other side and to promote their own
point of view. History has shown, however, that no matter what the
ideologues may believe and say, neither system is socially or economically
sustainable in its pure form.

While such negatives are not acknowledged in theory by the proponents of either system, they have long been recognized in practice. To
counter its own shortcomings, each system has adopted such features of
the other as are considered useful by its proponents. This has resulted,
as said earlier, in the creation of a variety of capitalist-socialist hybrids.
This process has smoothed the rough edges of either ideology, but has

not resulted in a satisfactory solution. The socialist components of the hybrid system, denounced as "big government" by free enterprise purists, tend to stifle the creativity of the free market. On the other side, the allowance made for free market incentives, which according to socialists are nothing but "greed," generates class distinctions and income inequality. We thus end up with both a bloated and expensive state structure and social inequality, combined within an economy that is neither very fair nor very efficient.

This unsatisfactory situation has its roots in the artificial antagonism that has been created for ideological reasons, and has pinned the state against the market. The capitalist camp's chief interest has always been primarily economic, mainly the preservation and increase of its wealth. In order to achieve this goal they needed to minimize all possible interference with their activities, particularly interference from the political power. The capitalist ideology tends, therefore, to reduce the role and power of the state and minimize its influence. The socialist camp, on the other hand, wanted from early on the overthrow of the existing order and its replacement by a new one that socialists would control. For this purpose they wanted state power to be as far-reaching and as complete as possible, while the influence of the markets was reduced or eliminated. Thus the state-market antagonism was born, and with it the two one-sided ideologies that can never truly coexist: on the one hand that state control over the economy should be as extensive as possible, on the other that the economy should be completely free of political interference. These two views have become the respective foundations of what we call today liberalism, which equals left, and conservatism, which equals right.

These irreconcilable views must now be abandoned because the economic situation on the basis of which they were initially developed begins to fade away. The challenge of the future is no longer to determine how apparently unlimited wealth, created by an industrial civilization based on cheap energy, will be distributed. The new goal is the preservation for the general welfare of some form of this wealth, and of the standard of living it makes possible. For this purpose the correct and

169

complimentary roles of the market and the state must be restated so as to eliminate the current antagonism, and replace this antagonism with cooperation.

Here we can, so as to provide concrete examples, go back to the WWII industrial mobilization effort. It is worthwhile noting that prior to the war the nation was sharply divided politically. The "anything goes" capitalism of the Roaring Twenties had led the nation into the Great Depression, with the resulting hardships spread wide among the general population. The Right had been shamed and the Left, clustering around FDR and the New Deal policies, was leading the nation on what could be called a socialist track. Republicans and Democrats cordially hated each other.

The ideological issues were pushed aside by the necessities of war. The U.S. needed to equip its forces with aircraft carriers, planes, tanks, ammunition, and myriad other items, in huge quantities. There was no doubt that only U.S. industry, the most inventive and dynamic in the world, could fulfill this monumental task. Ideology was set aside for the duration of the conflict. The Left did not talk about nationalizing industry, nor did the Right attempt to reduce government powers, which it now saw as justified by the national state of emergency. The government set the priorities and allocated resources, while industry produced, produced, and produced still more. It was in one sense a "command economy" but on the other hand initiative, inventiveness, and creativity flourished.

The United States' astonishing success in WWII demonstrated that the respective roles of the state and of private industry are not, by definition, antagonistic, but can be harmoniously combined for the pursuit of the national interest. The role of the state is to set policy within the framework of this interest as perceived by the population, and particularly so if the government is structured on a democratic model. The role of free enterprise is to provide and produce within that policy framework, and to ship the goods to wherever they are allocated by the policy makers. The government, by contrast, does not have to provide anything except strategic direction, a list of priorities and overall supervision. If that is clearly understood, then the state can be slimmed down to a lean minimum. The federal government, which ran the massive effort of WWII was in fact far smaller than the one we have today, yet it successfully staged and directed a national effort of immense complexity and magnitude.

What are the chances for a similar effort to be undertaken in the field of energy? In WWII both the challenge and the resulting focus were intense. The energy problem is somewhat more diffuse, and the resolution will take much longer than the three years and nine months it took the United States to win the Second World War. In terms of length, span, and intensity of effort, the energy task falls somewhere between WWII and the settlement of the West. America was successful at both, and there is no reason to believe we cannot do the same again.

Obviously a task of this magnitude requires a national consensus. On the surface there seems to be very little of that at the moment, but despite the appearances the main underlying situation is not without potential.

Two parallel trends can currently be observed in U.S. politics: a sharpening of ideological positions on the one hand, and a move to the center on the other.

INCREASED IDEOLOGICAL INTENSITY

There appears to be a sharpening of ideological differences. This was already apparent while President Clinton was in office and became even more pronounced under the Bush administration. The strong opinions and often polarized discourse heard during the 2008 election could be interpreted to mean that the dispute between the Right and the Left has taken on new and virulent life. If true, this would contradict our earlier statements about these ideologies becoming obsolete.

But such a conclusion is not inevitable. Ideology always takes on new life when unresolved issues accumulate, frustration grows and political

pressure rises. Ours is such a moment in time. The process of hybridization between socialism and capitalism has reached its final stage. As
mentioned earlier, the root cause of this blending process was not a
genuine search for a common solution, but an attempt by both sides to
compensate for their respective shortcomings without sacrificing their
fundamental ideological positions. The resulting situation can be lived
with, but is not satisfactory. Come a crisis or an unexpected challenge,
one must choose between the competing approaches: go Left or Right.
We have come just to such a juncture, and the old militant ideologies,
after a time of compromise and hybridization, are now being revived.

When a compromise breaks down, the parties return to their original
positions, sometimes with a vengeance, seeing, after a time of restraint,
a new opportunity to gain the ascendancy. The fact that the situation
has evolved under their feet does not touch them, because their respective ideologies is all they know, and they fall back on them as the only
acceptable solution. Thus we see the hard core of each party drifting
further to its respective end of the political spectrum.

As such, an increase in ideological commitment and intensity bodes
ill for the arising of a national consensus. There is, however, a countervail-
ing phenomenon also taking place: a general move to the center.

THE GROWING POWER OF THE INDEPENDENTS

The abysmal approval ratings of both Congress and President Bush
before the 2008 election show that the hard core of either party is far
from controlling the situation. Indeed as the committed base on each
side circles its ideological wagons, a much greater number of voters
move away from the extremes on either side and congregate in the
middle. This growing crowd of independent voters can be looked at
from two angles.

The conventional view is that they are either "moderates," lacking
any strong conviction, or "cherry-pickers" who like some pieces of the
platforms of either party but not others, and would prefer to see a

platform blended and customized to their taste. The accepted political wisdom is to attract such voters by "widening the tent" so as to include some goodies for everyone. This is nothing but a modernized version of the Left-Right hybridization process, which may fool some but offers no definite solutions.

The more hopeful interpretation is to see the independents as realists. In other words they are voters who know or sense that key issues are not being addressed and that solutions to the problems that they see developing are not forthcoming. From that viewpoint, the growing mass of independents at the center has the greatest potential for action, initiative, and achievement, but, being realists, they will not move and will remain uncommitted until they can see a political platform that will actually do the job. They have been there and done that, and they know the old ideas will not work. They are waiting for new guidelines and strategies that will lead to real solutions.

The independent center is there for the taking by either party, and they could be the stuff of a new majority, for in all likelihood they outnumber the hard cores of both parties put together. But to catch this prize the traditional parties have to get out of their ideological straightjackets and offer real solutions, as opposed to just emotionally satisfying ideological statements. The party that succeeds in "cornering the center" will gain the dominance, while the other party is most likely to fall to minority status for a generation or more. Whether either of the established parties is at this point capable of such internal flexibility and coherence is an open question.

A HISTORICAL PARALLEL

There is also a third possibility. Even as the mass of uncommitted voters is large enough to provide one of the existing parties with an overwhelming majority, it is also sufficient to support a new political formation. The current conventional wisdom is that such a hypothetical "third party" could possibly play the role of a spoiler, but that it cannot provide a credible challenge to the existing political establishment.

173

The chief argument for this view is that the financial and organizational resources needed to establish a credible new party would be so large as to make the creation of a genuine third party a practical impossibility.

In order to provide perspective in this respect, it is worthwhile to look at the historical record. The U.S. has always had, and probably always will have, a two-party political system. But this system does not exclude shifts in allegiance and transfer of loyalties from one faction or party to another. Historically, the appearance of such new political formations tends to cluster around times of frustration and stress, such as is now the case. "Third parties" tend to arise when, due to fundamental changes in the political and/or economic situations, the interests of a large group of voters are not adequately addressed by the powers that be.

Such a situation presented itself in the decade preceding the Civil War, and a cluster of third parties appeared at that time. Nearly all had a short life, but one, the newly founded Republican Party, was phenomenally successful and swept the nation. Formally established as a party in 1854, the Republicans were in control of both Congress and the White House by 1860, and after the Civil War remained the dominant political power for half a century.

The extraordinary success of the Republicans had a double foundation. The first pillar of their popularity with the voters was the party platform. Instead of protecting the status quo, it offered imaginative and practical solutions to the problems facing a nation that was both beginning to industrialize and was simultaneously expanding westward. These issues, with the solutions offered by the Republicans, were:

- Transportation links with the West: to encourage railroad investment, valuable land grants along the lines were provided.

- To provide technical cadres such as surveyors, engineers and veterinarians, colleges financed with land grants were to be established.

- A high customs tariff was established to protect U.S. industries against foreign competition.

- To satisfy abolitionists, the Republicans proposed to abandon the numerical parity between free and slave states.
- The Homestead Act provided free land to westward settlers and new immigrants.

There were also financial reforms favoring the rise of the emerging industrial capitalism.

These policies were aimed primarily at two new and growing constituencies: the western settlers in what has since become the Midwest, and the nascent industrialists of the Northeast. Neither of these constituencies could fit comfortably in either of the established parties, whose leaders were drawn from the traditional merchant and landowner elites of the eastern seaboard. At the same time these new and dynamic groups of voters yearned for a clear vision of where the nation was headed and of what was needed for it to get there. In addition, the Republican position on slavery gave the party the moral high ground, even though the Republicans were originally far from proposing outright abolition or emancipation. In this manner the Republican Party offered a clearly marked and accessible road to the future, leaving behind what was a deadlocked and somewhat corrupt political system.

The other pillar of the sweeping Republican takeover was their membership policy. They invited and welcomed into the new party any capable politician who agreed with their basic platform or a significant piece of it, regardless of previous political affiliation. Once the attractiveness of the proposed policies became evident, many able and experienced governors, congressmen, and state officials joined the party, bringing with them voter recognition, competence, and drive. Abraham Lincoln, a former Whig, was the top Republican prize, but there were many others, without which the party could never have achieved its phenomenal growth in numbers, power, and influence.

History does not repeat, but its lessons are useful, because people tend to respond in similar manner to the same kind of situations. There are several parallels between our decade and that of the 1850s. Then as now, the nation was faced with fundamental issues that the existing political establishment was, as a whole, unwilling or unable to tackle. Then as now, there were a large and growing number of uncommitted, disaffected, and "independent" voters whose interests and needs were not being addressed or even recognized. Yet this very group of neglected voters included some of the most dynamic and competent elements of the nation: the risk takers, the settlers, the industrialists, and the immigrants. All of those believed in the nation's potential and wanted it realized. It is these very people who not only fought in, and led, the Union armies in the Civil War, but also who, once the conflict was finished, went on to build modern America.

There is a similar and unrecognized potential in today's American electorate. The nation is on the verge of a major industrial and economic transformation, and is also laden with liabilities connected with the now passing age of continuous economic growth and its associated adversarial politics. There is general dissatisfaction with the direction in which the nation is currently led: a government that combines bureaucratic bloat with virtual paralysis; an economic agenda the impact of which on the general population is at best dubious, at worst downright disastrous; a political leadership ensconced in its own interests and privileges and incapable of offering anything except "more of the same"; and permeating it all, a general perception of a degradation of the quality of life, a sense that for the first time in U.S. history the next generation will have to do with less, in every area, than the preceding one: a hint of a loss of the American Dream.

There are two ways to react to such a situation. One is to remain on the same track, apply the same worn-out principles, and try to fix the situation in the same way as was done in the past. But as things do not work as they once did, this approach requires ever greater effort and expense in order to obtain constantly diminishing returns. This leads to disappointment, frustration, and eventually conflict, as is always the case when reality contradicts ideology, because ideology, by its very nature, requires an enemy to explain away failure.

The other approach is what we could call "the new realism": to recognize the reality of the change that is taking place and of its consequences, and to align our response accordingly. The previous chapters of this book have outlined what this response could be in the immediate future. This would set the stage for a longer-term national effort to deal with our energy situation. This issue is fundamental because fossil energy plays a greater role in America than in any other region or nation. We are, after all, by far the world's greatest user of energy on a per capita basis. Therefore, a shortage of such energy, beginning with oil, will affect us more than any other nation.

Such a change does not need to be catastrophic. It will be that only if we refuse to adapt and insist on carrying on as before despite circumstances that will make the "status quo" option an impossible one in the end. We have a choice: to try to remain what we now are, or re-invent ourselves for the dawning future. America is not simply a material entity, although the material aspect of our lives, the consumer society, is probably the most conspicuous one at this time. Regardless of its current condition, the United States has been conceived on a much higher plane than the purely material one. It is that higher definition of the nation that allows us to choose the way we will take, and to succeed in the endeavor.

NEXT STEPS

If you like what you have read, you may want to continue the conversation.

Join us at www.ViableEnergyNow.com to find more information, an opportunity for sharing ideas and a framework for action.

ABOUT THE AUTHOR

Jacek Popiel was born in Poland and educated in Africa, Canada and the United States.

After working as a teacher in Africa he joined the armed forces of Belgium as an infantry officer. As part of his service he spent two years in the United States, working in strategic analysis and planning for the U.S. Army.

After leaving the armed forces he worked for thirty years in international business development for a number of European and American companies. His areas of activity have included Western and Eastern Europe, Russia, North America and Japan. His most recent position involved automotive engine technology, and in that capacity he worked extensively with world-class automotive manufacturers as well as the U.S. Department of Energy and the National Renewable Energy Laboratory.

Jacek and Joanna Popiel have four grown children and live in Colorado Springs, Colorado.

www.ingramcontent.com/pod-product-compliance
Lightning Source LLC
Chambersburg PA
CBHW072141270326
41931CB00010B/1845